TELECOMMUNICATION WIRING

Clyde N. Herrick
C. Lee McKim

PRENTICE HALL
Upper Saddle River, NJ 07458

Library of Congress Cataloging-in-Publication Data

Herrick, Clyde N.
Telecommunication wiring / Clyde N. Herrick, C. Lee McKim.
  p. cm.
Includes bibliographical references and index.
ISBN 0-13-151531-4
1. Telecommunication—Equipment and supplies. 2. Electric wiring.
3. Electric apparatus and appliances—Installation. I. McKim, C.
Lee. II. Title.
TK5103.H47 1992
621.382—dc20   91-12165
                                                          CIP

Editorial/production supervision
  and interior design: Laura A. Huber
Production assistant: Jane Bonnell
Acquisitions editor: George Z. Kuredjian
Editorial assistant: Barbara Alfieri
Artist: Robert Mosher
Cover designer: Ray Lundgren Graphics

Copy editor: Sally Ann Bailey
Marketing manager: Alicia Aurichio
Prepress buyers: Kelly Behr/Mary
  Elizabeth McCartney
Manufacturing buyer: Susan Brunke

© 1992 by Prentice-Hall, Inc.
A Simon & Schuster Company
Upper Saddle River, New Jersey 07458

The publisher offers discounts on this book when ordered in
bulk quantities. For more information, write: Special
Sales/Professional Marketing, Prentice Hall, Professional &
Technical Reference Division, Englewood Cliffs, NJ 07632.

Printed in the United States of America

10 9 8

ISBN 0-13-151531-4

PRENTICE-HALL INTERNATIONAL (UK) LIMITED, *London*
PRENTICE-HALL OF AUSTRALIA PTY. LIMITED, *Sydney*
PRENTICE-HALL CANADA INC., *Toronto*
PRENTICE-HALL HISPANOAMERICANA, S.A., *Mexico*
PRENTICE-HALL OF INDIA PRIVATE LIMITED, *New Delhi*
PRENTICE-HALL OF JAPAN, INC., *Tokyo*
SIMON & SCHUSTER ASIA PTE. LTD., *Singapore*
EDITORA PRENTICE-HALL DO BRASIL, LTDA., *Rio de Janeiro*

Contents

PREFACE ix

CHAPTER 1 ELECTRICAL CHARACTERISTICS OF WIRE 1

    1.1   Introduction 1

    1.2   Voltage in an Electrical Circuit 2

    1.3   Current in an Electrical Circuit 3

    1.4   Resistance in Wiring Circuits 4

    1.5   Power and Power Loss 6

    1.6   Signal-to-Noise Ratio 6

    1.7   Inductance and Inductive Reactance in Wiring Circuits 7

    1.8   Capacitance and Capacitive Reactance in Wiring Circuits 8

    1.9   Impedance in Wiring Circuits 9

    1.10   Digital Signals 12

    1.11   Analog Signals 13

    1.12   Ground and Grounding 15

    1.13   Cross-Talk in Wiring 17

    1.14   Attenuation of Signal Information 17

1.15   Insulation of Conductors   18

Summary   22

Questions   22

CHAPTER 2   TRANSMISSION MEDIA TWISTED PAIR   **23**

2.1   Introduction   23

2.2   Twisted-Pair Cabling   24

2.3   Cross-Talk on Twisted-Pair Cable   25

2.4   Shielding of Transmission Lines   26

2.5   Applications and Functions of Twisted Pairs   28

2.6   Special Applications of Twisted Pairs   31

2.7   Twisted-Pair Cable Termination   31

2.8   Distribution Frames   36

2.9   Existing Cable Systems and Compatibility   37

2.10   Electrical Characteristics of Twisted-Pair Cables   40

2.11   Tools for the Installation of Twisted-Pair Cables   43

2.12   Advantages of Twisted-Pair Cabling   46

2.13   Disadvantages of Twisted-Pair Cabling   47

Summary   48

Questions   48

CHAPTER 3   COAXIAL CABLE   **49**

3.1   Introduction   49

3.2   Characteristics and Construction   50

3.3   Coaxial Cable Connectors and Terminations   53

3.4   Grounding of Coaxial Cables   64

3.5   Applications of Coaxial Cables   68

3.6   Advantages of Coaxial Cables   68

3.7   Disadvantages of Coaxial Cables   70

Summary   71

Questions   71

CHAPTER 4   FIBER-OPTICS   **72**

4.1   Introduction   72

4.2   Fiber Types   74

4.3   Cable Construction   76

4.4   Cable Characteristics   77

4.5   Cable Terminations   81

4.6   Preparing a Connection and Termination   85

4.7   Advantages of Fiber-Optic Cabling   89

Summary   92

Questions   93

CHAPTER 5   BASIC NETWORK TOPOLOGIES   94

5.1   Introduction   94

5.2   Basic Network Models   94

5.3   Determining Network Connections   95

5.4   Point-to-Point Topologies   96

5.5   Multipoint or Multidrop Networks   96

5.6   Bus Network   97

5.7   Star Topology   99

5.8   Hierarchical Network   102

5.9   Ring Network   102

5.10   Network Access Protocols   102

Summary   105

Questions   105

CHAPTER 6   PLANNING THE WIRING INSTALLATION   106

6.1   Introduction   106

6.2   Project Scope   107

6.3   Existing Cabling   107

6.4   User Population   108

6.5   Number and Type of Work Areas   108

6.6   Documentation and Room Layout Database   108

6.7   Type and Number of Devices Required   109

6.8   Phones and Station Equipment   110

6.9   Maximum Power Allocation and the Number of Power
           Outlets   110

6.10   Service Areas Affected by Work to
           Be Performed   111

6.11   Review of Building Plans and Cable
           Requirements   111

6.12   Test Equipment and Commitment to Support
           Personnel Training   112

6.13   Telco, Voice, and Data Support Room
        Requirements   112

6.14   Environmental Concerns   114

6.15   Grounding and Bonding   115

6.16   Cable Network Mechanical Supports   116

6.17   Electromagnetic Interference   116

6.18   The Question of User Device Ownership   116

6.19   Hot Host Service   117

6.20   Building, Office, and Device Inventory   118

6.21   Network and Plan Documentation   120

6.22   Quality and Electronic Control   122

6.23   Service Impact Severity Classifications   122

6.24   Plan Review by All Affected Parties   123

6.25   Establish a Service Desk   124

6.26   Scheduling the Job   126

6.27   Writing the Request for Bid Proposal   126

6.28   Documentation Responsibility   127

6.29   Installation of the Wiring   127

6.30   New Building Application   127

6.31   Establishing a Labeling Scheme   128

6.32   Database Tracking System   128

6.33   Safety   129

Summary   129

Questions   129

CHAPTER 7   INSTALLING THE CABLE   **130**

7.1   Introduction   130

7.2   Making the Wiring Plans   131

7.3   Cable Strategy   132

7.4   Two-Point Connection Strategy   132

7.5   Three-Point Connection Strategy   133

7.6   Four-Point Connection Strategy   135

7.7   Rules for Installing Cables   135

7.8   Installation Techniques   137

7.9   Cable Installation Hardware   147

7.10   Grounding the Cabling System   155

Summary   155

Questions   156

CHAPTER 8   TESTING AND TROUBLESHOOTING   158

8.1   Introduction   158
8.2   Objectives of Testing and Troubleshooting   158
8.3   Testing Twisted-Pair Wires   161
8.4   Continuity Test of a Cable   161
8.5   A Short to Ground Test   163
8.6   Testing a Coaxial Cable   164
8.7   Testing a Fiber-Optic Cable   169
8.8   Custom-Assembled Cables   169
8.9   Signal-to-Noise Ratio Measurements   169
8.10   Reference Point for Power Level   172
8.11   Zero Transmission Level   173
8.12   The Technical Support Center   175
Summary   175
Questions   175

CHAPTER 9   DOCUMENTATION OF THE WIRING SYSTEM   176

9.1   Introduction   176
9.2   Labeling the Cabling System   177
9.3   Blueprints and Diagrams   178
9.4   Distribution Logs   178
9.5   Work Area Inventory Sheets   179
9.6   Handwritten Entry Versus Terminal-Based Entry   179
Summary   182
Questions   182

CHAPTER 10   TELECOMMUNICATION DATABASE   183

10.1   Introduction   183
10.2   File-Based Tracking System   184
10.3   Data-Based Tracking System   185
10.4   Structured Query Language   185
10.5   Basic Components of a Database
         Management System   186
10.6   Database Manager's Responsibility   187

10.7   Sample Database   187

10.8   Field Description of Master Panel   189

10.9   Type of Request Section   192

10.10   Field Description for Link and Connect Panel   195

10.11   Service Availability Panel   197

10.12   Field Descriptions for the Work Order Panel   197

Summary   201

Questions   201

**CHAPTER 11   MANAGEMENT AND THE WIRING
MANAGEMENT PROBLEMS   202**

11.1   Introduction   202

11.2   Tracking   204

11.3   Measurement, Testing, and Troubleshooting   204

11.4   Retrofitting   204

11.5   Cost Factors   204

11.6   Database and Database Development   204

11.7   Writing a Bid Proposal Request   205

Summary   206

Questions   207

**CHAPTER 12   WRITING THE SPECIFICATIONS FOR A
BID PROPOSAL   208**

12.1   Introduction   208

12.2   Details to Include in a Request for Bid Proposal   208

12.3   Development Time   209

12.4   Special Contractor Considerations   209

12.5   Bid Proposal Forms   210

12.6   Detailed Work Proposal   210

Summary   226

Questions   226

GLOSSARY OF TERMS   227

REFERENCES   241

APPENDIX: PERIODICALS   243

VENDOR INFORMATION   244

INDEX   247

Preface

The reasons for this book are many. Some are obvious to the telecommunication manager and department personnel. Others are not so obvious. The data communications field is changing rapidly but the way that business operates is being challenged at every corporate level. Now more than ever resources, including wiring systems, need to be utilized to their maximum.

The need for special wiring systems and greater capacity cabling for data communications equipment has created the generation of new job categories in the workplace such as: telecommunications manager, communication wiring planner, information systems manager, connectivity specialist, communication wiring technician, and so on. In our occupational area, it is obvious that there is a need for a text dedicated to the "nuts and bolts" of telecommunication wiring systems and cabling.

Many books have been written on higher level subjects in telecommunications such as: local area networks, designing LANs, telecommunications systems, and so on. However, the cabling and wiring sections of such books seldom offer any practical information for those involved in the designing, installation, testing, or updating of wiring systems that are critical to the operation of any telecommunication system. The cabling should be treated as a "dynamic resource" rather than a static one. These systems, whether a single coax or a complete wiring plant, should be treated as a major support sub-system.

Management will find helpful the discussion on the importance of having a complete inventory of installed cable and wiring runs to determine "in place-capacity'' versus ''in-placed used capacity.'' The chapter on task management will assist the manager in giving direction and leadership to the installation team, the maintenance team and upper management. The section on writing a bid proposal can be helpful to management in preparing the proposal and evaluation of the finished product.

Telecommunication cable installers, planners, managers, and audit teams should find useful the discussion on standardization in setting up methods for identifying and labeling. This topic will be particularly helpful if the system has gone through several installs without a set of universal standards. The suggested standards should be a help in the establishment of corporate labeling standards for cabling, patch panel, wiring closets, floor locations, and equipment.

The wiring specialist and telecommunication planners/designers should find the topics on cabling systems, supports and test hardware, proper installation techniques, and wire and fiber characteristics useful in the planning of a cabling system. Finally, the chapter on planning and wiring installation offers the wiring specialist guidelines for planning, installation, and testing of the cabling system.

Our attempt is to establish a reference point from which logical decisions in the designing of a cabling system, selection of the media type, writing the job proposal, documenting systems and establishing a maintenance facility can be completed. We fully understand that every company has unique telecommunication needs and that every wiring system will be different.

When the text inclusions contain certain trade names or trademark items, this is not to be taken as an endorsement by the authors of any particular product. These illustrations are included to illustrate to the reader some of the more successful products and alternatives available on the market today. There are many manufacturers and vendors for any of the items mentioned, and it is the responsibility of the professional to keep abreast of the literature. To this end, the authors have included names and addresses of many of the periodicals in the appendix, along with the names and addresses of the companies whose products are mentioned in the text.

The authors attempted to make the material in the text is as ''state of the art'' as possible, fully realizing that technology changes daily in this field. Telecommunication invocation is perhaps the fastest growing technology in the world today. Cabling systems will be static neither in the media in which it is used nor in the types and quantity of data that they will transmit.

Finally, the authors wish to express their appreciation for all the companies and individuals who have supplied information for the development of this book.

Clyde N. Herrick
C. Lee McKim

CHAPTER 1

Electrical Characteristics

1.1 Introduction

It may not be necessary for the worker who installs wiring systems to understand the electrical properties of wires and conductors to install a system correctly. However his or her understanding of these properties will give a better appreciation of the job to be done and the tools that are to be used.

It is necessary for the staff person responsible for the telecommunication system and for the wiring system designer to understand the electrical properties of wiring and how their presence in cables and wires affects the signal information. The purpose of this text is to assist the cable installer and the wiring system designer.

A wiring system is a form of an electrical circuit, an electric circuit that is comprised of an energy source, an energy transfer medium, and a load.

An **energy source** could be a battery, a generator, an amplifier, a digital computer, or any of the other devices that output energy in form of a voltage and current.

An **energy transfer medium** is any material used to transport energy from one place to another. Transfer media could be copper wires (conductors), fiber-optic cables, or the air in the case of radiated energy.

A **load** in a circuit can be any of many components or devices that receive the transferred energy such as: resistors, light bulbs, speakers, motors, computer terminals, printers, PCs, and so on.

To understand the concept of a circuit better, energy transfer, sources, and loads, we must introduce the concepts of voltage, current, resistance, power, and energy transfer.

1.2 Voltage in an Electrical Circuit

An **electromotive force** (EMF) is that force that tends to move electrical energy. EMF or voltage is conveniently regarded as an attractive or repulsive force measured in **volts.** Voltage can be compared to water pressure that causes water flow in a pipe which is measured in pounds per square inch (psi).

A voltage is a difference of potential or electromotive force that attracts and repels electrons. Another way of thinking of voltage (V or E) is a force or pressure that forces electrons through the circuit. It is this movement of electrons that carry the energy throughout the circuit.

DC voltage (direct-current voltage) is the name given to voltage in a circuit where the current flows in only one direction and is either positive or negative. This usually means that it is a value above ground (positive) or below ground (negative).

Ground potential or **reference** is thought of as 0 volts. Ground potential is the "potential of the earth" (in England, the term is "earth"). The term "ground" is also used to mean the metal case or chassis of a piece of electronic equipment. We will discuss this in more detail later in the chapter.

The polarity of a voltage is usually discussed in reference to ground. A value above ground, for example, 10 volts, is said to be +10 volts, while a voltage of 10 volts below ground is said to be −10 volts.

A battery such as that shown in Figure 1-1 has a positive and a negative terminal. The positive side is +9 volts in respect to the negative side. On the other hand, the negative side is −9 volts in respect to the positive side. When a battery or other dc voltage is connected in a circuit, a dc current (electrons) flows from the negative terminal of the battery through the circuit and returns to the positive terminal. This theory of current flow is called **electron flow.** We should note here that most engineers prescribe to positive current flow or conventional current flow simply because it is **conventional,** that is, was the first theory of current flow. The end results are the same; that is, energy is transferred through the circuit. In electrical circuits, a battery voltage or supply voltage is denoted as

Figure 1-1 A 9 volt battery.

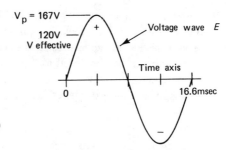

Figure 1-2   A 120 volt 60 Hertz (Hz)
sinewave.

E while voltage drops in a circuit are symbolized as **V.** Voltages can be developed
from many sources, such as batteries, solar cells, thermocouples, generators, and
so on.

AC voltages (alternating-current voltages) vary above and below ground
with respect to time. An example of an ac voltage is common house voltage that
changes at a rate of 60 cycles per second (hertz). Figure 1-2 illustrates a cycle of
voltage from a 60 H source. AC voltages change with time and are also called
analog voltages.

In this text we are interested in analog voltages and digital voltages. As we
stated earlier, analog voltages are those that vary with time, such as voice signals
(Figure 1-3a). Digital signals are in form of pulses or bits (Figure 1-3b) that
change quickly from one level of voltage to another.

1.3 Current in an Electrical Circuit

Current or **current flow** is a movement or flow of electrons through a circuit that
is caused by the electromotive force or voltage applied to the circuit. When a
voltage is applied to a complete circuit, from a source, electrons move through
the conductors of the circuit (energy transfer) to the load at the receiving end of
the conductors. The desire is to transport the electrical signals along the conductors (wires) and have them arrive at the destination in the same configuration and
with the same voltage level as when they left the source. In other words, the system
must reproduce the input signal at the receiver to relay the input information correctly, be it voice signals, other analog signals, dc voltages, or digital pulses.

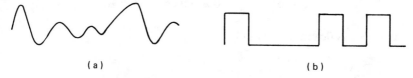

(a)   (b)

Figure 1-3   Examples of voltages:
a. analog voltages.
b. digital voltages.

We should note here that fiber-optic cables transport energy via photons of light energy. However, the devices that produce the signals and the devices that receive the signals depend upon electrical energy for operation. The signal from the source device must be converted to light energy to be transported by the fiber cable and then converted back to electrical energy to be used by the destination device. Fiber optics will be discussed in some detail in Chapter 4.

1.4 Resistance in Wiring Circuits

Resistance is the property of an electrical circuit that limits the current. Resistance can be compared to friction in a mechanical device where the frictional drag on a body limits the speed of the body and produces heat. The resistance of a material limits the number of electrons that flow in the material—the amount of energy that is transferred—and causes heat in the circuit. Resistance occurs in all electrical circuits; even the best conductors have some resistance.

The unit of resistance is the **ohm** and is symbolized by the Greek letter omega (Ω). The relationship between the voltage, current, and resistance in a circuit is called **Ohm's law** and is formulated by

$$\text{amperes} = \frac{\text{volts}}{\text{ohms}}$$

$$I = \frac{E}{R}$$

Ohm's law states that current is directly proportional to voltage and inversely proportional to resistance. That is, increased voltage causes increased current flow, and increased resistance decreases the amount of current.

Figure 1-4 depicts a simple circuit of a battery and a light and the schematic diagram of the circuit. The schematic diagram of the circuit assumes that the wires to and from the 50 ohm load have no resistance.

The current is determined by Ohm's law:

$$I = \frac{E}{R}$$

$$I = \frac{10 \text{ V}}{50 \text{ ohms}} = 0.2 \text{ amperes or } 0.2 \text{ A}$$

The resistance is determined as follows:

$$R = \frac{E}{I}$$

$$R = \frac{10 \text{ V}}{0.2 \text{ A}} = 50 \text{ ohms}$$

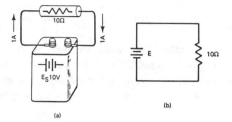

Figure 1-4   A simple electrical circuit:
a. circuit.
b. schematic diagram of the circuit.

We noted earlier that conductors (wires) are not perfect energy transporters but have resistance. In most cable runs we can ignore wiring resistance, but in long cable runs we might experience situations as depicted in Figure 1-5. The 1 ohm resistors between the source and the load represent the wiring resistance. The current in this circuit is determined as follows:

$$R = R1 + R2 + R\ \text{load}$$

$$R = 1 + 1 + 50 = 52\ \text{ohms}$$

The current is

$$I = \frac{E}{R} = \frac{10\ V}{52} = 0.0192\ \text{amperes}$$

The load voltage is:
$$\text{load} = I \times R\ \text{load} = 0.192 \times 50 = 9.6V$$

There is a loss of 0.4 volts along the line.

$$\text{signal loss} = R\ \text{wire} \times \text{current}$$

$$SL = 2\ \text{ohms} \times 0.0192\ A = 0.4\ \text{volts}$$

This loss of signal voltage can also be related to a loss of signal power as we will see in Section 1.5. If the length of the wires in the foregoing example were

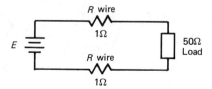

Figure 1-5   An example of an electric
circuit with wiring resistance.

doubled, the loss would double. Obviously, very long lines would cause excessive signal loss.

The resistance of wires is determined by both length and cross-sectional area. The smaller the cross-sectional area of the conductor, the greater the resistance for a given length, and the longer the conductor, the greater the resistance for a given cross-sectional area. The properties of wire will be discussed further in Chapter 2.

1.5 Power and Power Loss

The primary purpose of transmission lines regardless of the type is to transfer energy or power from one device to another. Power is the time rate of doing work. In electrical circuits the power is the **watt** and is formulated:

$$\text{power (in watts)} = \text{current (in amperes)} \times \text{volts (in volts)}$$

$$\text{or power} = I^2 \times R \text{ (watts)}$$

Power is the unit that we most often use when relating to the levels of signal in a circuit. We usually refer to the power loss in wires as $I^2 R$ losses.

Table 1-1 presents a summary of the formulas for both Ohm's law and the electrical power laws for resistive circuits.

1.6 Signal-to-Noise Ratio

Noise is the introduction of any unwanted signal into the system. Noise may be in the background of an audio signal and is an audio signal other than the desired

TABLE 1-1 A Summary of Ohm's Laws and Power Laws for the Resistive Electrical Properties in Circuits

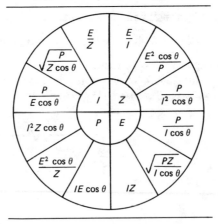

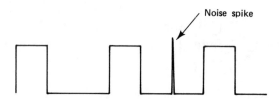

Figure 1-6   An example of noise spikes introduced into a digital pulse stream.

signal. In the case of digital signals noise may appear as analog signals or spikes that can mimic the digital pulses. Figure 1-6 illustrates the introduction of noise spikes into a digital pulse train. These noise spikes can be interrupted by a micro-processor or printer as digital pulses that represent a signal code other than the desired one. Although most systems contain a parity check (bit count), the intro-duction of two noise spikes would not necessarily be detected.

In either an audio or digital system there is a threshold level below which noise can be tolerated without interference or damage to the outcome of the sys-tem. The ratio of signal to noise *(SN)* in any system is formulated:

$$SN = \frac{\text{signal power}}{\text{noise power}}$$

For example, if an audio system had a signal level of 100 mW and the noise level was 2 mW, the *SN* ratio would be 50 to 1.

1.7 Inductance and Inductive Reactance in Wiring Circuits

Inductance *(L)* is the property of a circuit that causes an opposition to any change of current within the circuit. As electrons (current) move through a conductor, a magnetic field is produced. This magnetic field induces a voltage in the in-ductor, called a **counterelectromotive force (CEMF),** that opposes the current flow in the circuit. When a current flow tries to decrease in an inductor the CEMF is also produced that tries to keep the current flowing in the circuit. CEMF can be thought of as inertia that tries to prevent change. In other words, the in-ductor reacts in response to a current change. Therefore, we call this phenome-non **reactance.**

Reactance caused by the change produces a voltage that opposes the source voltage which is producing the change. This induced voltage is formulated as:

$$V_{\text{induced}} = L \left(\frac{di}{dt}\right) \text{ volts}$$

The formula states that the induced voltage *(V)* is equal to the inductance of the coil in henries times the change in current *(di)* over the change in time *(dt)*. The unit of inductance is the henry. If a current change of 1 ampere in 1 second produces an induced voltage of 1 volt an inductor has an inductance of 1 henry.

(a)   (b)

Figure 1-7   An example of an inductor and the symbol for an inductor.

Let us now put the inductive effect in context as to its effects on a digital circuit. When the source voltage (say, the output of a PC) increases, the inductive reactance causes a countervoltage that slows down the voltage change at the load terminals (of, say, a printer). Conversely, inductive reactance would slow down a decreasing voltage. This property of inductance can cause severe distortion in digital signals where the bits must change from zero to maximum voltage and from maximum voltage to zero voltage in short periods of time, often less than a millionth of a second. The **inductive reactance** of an inductor is opposition that an inductor offers to a changing current and is formulated as:

$$X_L = 2\pi fL$$

Inspection of the formula reveals that higher frequencies result in greater values of reactance and, therefore, that high-frequency circuits experience more signal loss than low-frequency circuits.

Wire has a small amount of inductance per meter of length; however, the inductance increases as the wire length increases, much like wiring resistance. The symbol for inductance is shown in Figure 1-7.

Cross-talk, the introduction of signals between conductors in close proximity to each other, is the result of the electromagnetic flux lines that are caused by the signal currents flowing in a conductor. **Flux lines** are invisible magnetic lines of force that are produced by the current (electron flow) in a circuit. The noise introduced from these flux lines, called cross-talk, can cause error signals to be introduced in a data line and unwanted conversation noise in audio lines. We will discuss the methods utilized to reduce cross-talk later in this chapter and in the chapters on cabling.

1.8 Capacitance and Capacitive Reactance in Wiring Circuits

When two metals are separated by an insulator, a **capacitor** is formed; the symbol for a capacitor is shown in Figure 1-8. Capacitors have the ability to store energy in the form of an electrostatic charge.

When one plate of a capacitor has more electrons (negative charge) than the other plate, a difference of potential or charge exists between the plates through the insulation. This charge is the result of the force from the electron on one plate

Figure 1-8   An example of a capacitor and the circuit symbol for a capacitor.   (a)   (b)

acting on the electrons on the other plate. The insulator between the plates is called a **dielectric.** The charge that is stored between the plates tends to oppose any change in circuit voltage. This opposition to a change in voltage is called **capacitive reactance.**

A capacitor is formed by any two conductors separated by an insulator. This means that there is a capacitance between a pair of conductors in a cable, a conductor and a ground, or a conductor and a shield. The reactance of a fixed capacitor or wiring capacitance is formulated as follows:

$$X_C = \frac{1}{2\pi FC}$$

where $\pi = 3.14$
F = frequency of the signal
C = capacitance of the capacitor

A capacitor has a capacitance of 1 **farad** when a current of 1 ampere causes a voltage change, across the capacitor, of 1 volt in 1 second.

All wires have resistance, capacitance, and inductance in varying amounts. Either or all of these factors can cause attenuation and deterioration of a pulse or an analog signal. Different types of lines have different amounts of these three factors. However, it is the resistance and capacitance that result in most of the losses in transmission lines and the inductance that results in the pickup of noise.

The longer the lines, the greater the amounts of these three factors: resistance, inductance, and capacitance. The increased amounts of these three factors result in increased deterioration of digital signals and increased analog signal loss.

1.9 Impedance in Wiring Circuits

The current and voltage in a resistor are always in phase with each other. That is, a maximum voltage results in a maximum current, and a minimum voltage follows a minimum current. On the other hand, the current and voltage in an inductor and a capacitor are 90 degrees out of phase with each other. The current in a capacitor *leads the voltage by 90 degrees* and the current in an inductor *lags the voltage by 90 degrees.* An example of this phenomenon is depicted in Figure 1-9.

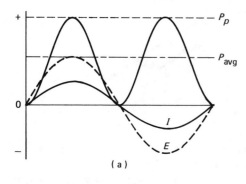

(a)

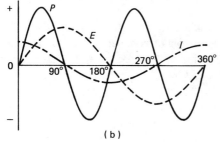

(b)

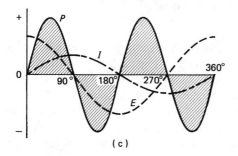

(c)

Figure 1-9 The relationship between current and voltage in an AC circuit:
a. in a resistor
b. in a capacitor
c. in an inductor.

The voltage drops around the circuit with resistance, capacitance, and inductance in Figure 1-10 is written as:

$$E = \sqrt{V_R^2 + (jV_L - jV_C)^2}$$

The $+j$ and $-j$ mean $+90$ degrees and -90 degrees, respectively.

Impedance is the name given to the total opposition to the flow of electrical energy in a circuit and is the result of a combination of resistance, inductance, and capacitance. The symbol of impedance is Z, and the unit of impedance is the ohm. The formula for impedance is

$$Z = \sqrt{R^2 + jX^2}$$

The jX indicates that reactance must be treated differently from resistance. Capacitive reactance is denoted as $-jX_C$ and inductive reactance is denoted as

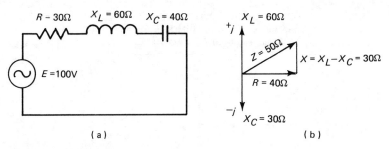

Figure 1-10   An electrical circuit with resistance, inductance, and capacitance:
a. circuit diagram.
b. vector diagram of voltages.

$+jX_L$. As was mentioned earlier, the $+j$ and the $-j$ can be considered to indicate $+90$ and -90 degrees, respectively. This means that Z must be calculated as shown in Figure 1-11.

The Pythagorean theorem states that

$$Z = \sqrt{R^2 + X^2}$$
$$Z = \sqrt{(100)^2 + (100)^2} = 141 \text{ ohms}$$

Impedance is a rather complex concept in AC circuits, and it is not within the scope of this text to offer a complete study of the subject. If more information is desired, the authors suggest one of the several fine basic electronic fundamental texts available. For our purposes we need to understand only that any output device has impedance, that transmission lines have impedance, and that any telecommunication device that is a load has impedance.

Transmission lines have a characteristic impedance, and loads are rated at an impedance value. We will discuss the importance of matching impedance of the transmission line to the device impedances in Chapter 4.

Table 1-2 is a similar summary of the formulas for electrical properties in inductive circuits.

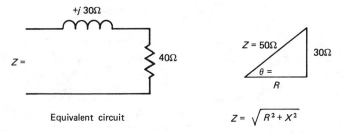

Figure 1-11   The impedance of a circuit must be calculated by using a right triangle and the Pythagorean theorem.

TABLE 1-2 The Presentation of the Electrical
Properties of Current, Voltage, Impedance, and
Phase Angle in AC Electrical Circuits

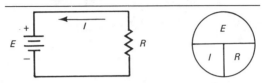

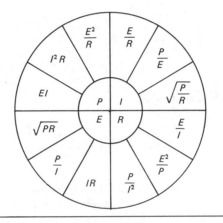

1.10 Digital Signals

Digital signals are discretely variable in the form of pulses. The pulses may represent a digital code that can be interpreted by a computer or other digital device as instructions, information, or data. Examples of instructions are add, save, or fetch. Examples of data are $+3$ volts, -30 degrees, or \$100.00. Examples of information are where to save (as an address) within the computer. Digital pulses are usually coded in a series of voltages and no voltages or ones and zeros, as shown in Figure 1-12. These digital pulses are called **bits.** A group of eight of these bits is called a **byte.**

The rate at which digital pulses are transmitted is called the **baud** rate. Baud rate is defined as bits per second or pulses per second and is directly related to frequency. The amplitude of the pulse is the negative or positive peak voltage as shown in Figure 1-13.

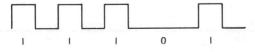

Figure 1-12 A digital pulse train comprised of voltages and no voltages representing ones and zeros.

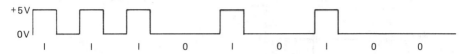

Figure 1-13 A digital pulse with a +5 volt amplitude.

A pulse often appears to rise and fall in zero time when observed on an oscilloscope; however, this is never the case. The capacitive reactance and the inductive reactance within the circuit cause the pulse in Figure 1-14a to actually appear as shown in Figure 1-14b. Each pulse has a rise time and a fall time as shown in Figure 1-15.

The **rise time** is measured between the 10% point and the 90% point, and the **fall time** is measured between the 90% point and the 10% point (Figure 15c). The zero and 100 percentage points are not used to measure the rise and fall times due to the difficulty of identifying their exact locations.

With very long transmission lines, where the inductance and capacitance are excessive, a pulse train may become so distorted as to become unrecognizable as shown in Figure 1-16. Special equipment can sometimes reconstruct the signal. Reconstruction of digital signals is much easier than reconstruction of analog signals. For this reason analog signals are often converted to representative digital signals before transmission and reconverted to analog signals at the receiver. Circuits that perform this function are called analog to digital converters (A to D) and digital to analog converters (D to A), respectively.

1.11 Analog Signals

Analog signals are any signals other than pulses. An analog signal has a voltage that is variable with time and is usually continually variable. Some examples of analog signals are shown in Figure 1-17. The signal does not have to vary at a constant rate.

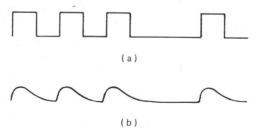

(a)

(b)

Figure 1-14 Pulse deterioration caused by the capacitance and inductance in the circuit:
a. pulse into the transmission line.
b. distorted pulse.

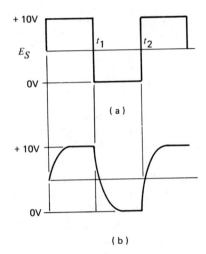

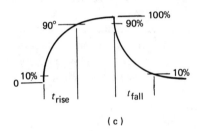

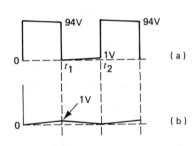

Figure 1-15 A typical digital pulse:
a. theoretical.
b. actual pulse shape.
c. rise & fall time.

Figure 1-16 Reactance in a circuit can make the pulse unrecognizable to the destination equipment.

Figure 1-17 Examples of analog signals:
a. a ramp signal.
b. a nonlinear analog signal.
c. a sinusoidal signal.

Examples of analog signals are voice or music (audio), sampling voltages (as from a pressure gauge), or dc voltages. Signals that have a periodic repetition rate (period) have a frequency. This repetition rate or frequency, in cycles per second, is called hertz or Hz. Frequency is formulated as

$$F = \frac{1}{T} \, \text{Hz}$$

1.12 Ground and Grounding

Ground, or earth as the British say, is usually referred to as the reference level for voltage levels within a system. U.S. government safety codes specify that all electrical equipment must be electrically connected to ground (grounded) to prevent an electrical potential from occurring between the equipment and ground and between equipments. Equipment grounding is a safety precaution designed to protect both people and equipment. The symbol for ground is shown in Figure 1-18. The chassis of most equipment is grounded and the return path for current flow in the chassis. This reduces the necessity for returned wires from the components. Most voltages in electronic equipment are measured to ground (the chassis).

Grounding is the connecting of an electrical circuit or equipment to the potential of earth. Grounding of electrical equipment is usually accomplished through the power plug. For 120 volt connections, the center prong of the three-prong plug is ground (Figure 1-19). The insulation color of the ground wire in a power lead to a piece of equipment is usually green or green and yellow.

Grounding of the shielding wire in telecommunication cable is important to assure the transmission of electrical signals along a cable without interference from the electromagnetic radiation from other transmission lines and electrical equipment. This interference, called **cross-talk,** can originate from adjacent lines, electrical motors, PCs, fluorescent lights, and so on. The term cross-talk originated from the phenomenon of the conversation of an adjacent telephone line being audible on the other.

While equipment grounding is primarily for the protection of people from electrical shock, there are other compelling reasons for grounding. Grounding provides a low-impedance path for electrical energy:

1. For protection for people from electrical shock in the event of an internal short in equipment
2. For protection of semiconductor devices from excessive static voltage buildup

Figure 1-18   The symbols for ground:
a. chassis.
b. earth.
c. common.

(a)   (b)   (c)

NEMA 5-15P

CONTACT VIEW

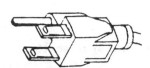

BACK

Figure 1-19 Examples of grounded ac power plugs:
a. three prong.
b. three-prong locking type.

3. To provide a safe path for electrical energy from lightning to protect both equipment and people

4. To provide a low-resistance path around the signal-carrying wires for low-frequency electromagnetic energy from sources such as power lines, light, motors, and so on.

5. To provide a low-resistance path around the signal-carrying wires for electromagnetic radiation from high-frequency electromagnetic waves from computers, other transmission lines, radiated signals (radio, TV), and so on.

The grounding system for a facility should maintain all the grounds of all telecommunication equipment, other electronic/electrical equipment, and all electrical power lines at the same potential, within closely prescribed limits of the National Electrical Code (NEC).

The **earth ground system** is the reference for all grounds within a building. The earth ground is established by the insertion of bronze or copper rods into the earth or copper wire into the concrete foundation of a building. This part of the ground system is the most difficult to establish to assure long-range effectiveness, because of the wide range of soil types and the varying moisture content of the soil. The moisture content of soil and the minerals within the soil determines the resistance of the soil, and thereby, the effectiveness of the ground in maintaining a low-voltage interface. Figure 1-20 depicts an example of a plant grounding system referenced from the NEC on proper grounding. Whenever possible, the connection to the ground electrode should be less than 1 foot below the surface of the soil, and the grounding electrode should extend at least 10 feet below ground. Ground in soil types other than moist clay requires special installation techniques. For example, grounding in shallow soil requires that grounding cable be laid in trenches and the soil compacted.

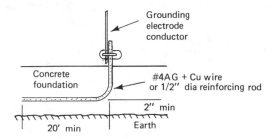

Figure 1-20   Example of a plant ground system.

Equipment and cable grounding will be discussed at the appropriate points within the text. All grounding installations must be in compliance with the latest edition of the National Electrical Code handbook (see NEC section 250).

1.13 Cross-Talk in Wiring

Electromagnetic pickup and radiation can cause serious problems in telecommunication systems, such as signal distortion, noise in conversations, and breach of security from a system. Some types of cables are protected from inductively induced signals (cross-talk) from adjacent lines. For example, pairs of wires are usually twisted to reduce inductive effects, and cables can be shielded from outside electromagnetic lines by surrounding the signal carrying wires with a braided or solid metal shield. There are also installation procedures that can reduce the effects of magnetic induction noise such as

1. shielding the cables
2. never running data cable in a conduit with power cables
3. using proper grounding of equipment and cables to protect against surge voltages and to provide shielding against outside signals

Grounding will be discussed in greater detail in Chapter 2 and the following cable chapters.

1.14 Attenuation of Signal Information

As was stated earlier wire has resistance, inductance, and capacitance. All these factors attenuate both digital and analog signals. The attenuation can be measured as either a reduction of voltage or a loss of power. This loss is usually referred to as a decibel loss. The bel is the logarithm of the ratio of power output to the power input of a system, in the case of a cable, the power into the cable at the

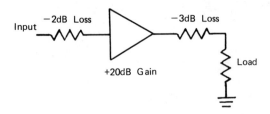

Figure 1-21   Example of dB gains and losses in an electronic circuit.

source and power output at the receiver. The bel unit is such a huge number that the decibel (1/10 bel) is usually used for calculations.

The formula for the decibel (dB) is

$$dB = 10 \log \frac{P_o}{P_{in}}$$

The decibel gain or loss can also be formulated using voltages:

$$dB = 20 \log \frac{V_o}{V_{in}}$$

For example, if a signal of 1 watt was put into a line with a resulting attenuation to 0.5 watt at the receiver end, the decibel loss is

$$dB = 10 \log \frac{0.5}{1} = 10 \log 0.5 = -3$$

We would say that the signal had a −3 dB loss.

As another example, suppose that a 1 volt signal were injected into a transmission line with a reduction of to 0.707 volts at the receiver end. The dB gain would be

$$dB = 20 \log \frac{0.707}{1} = 20 \log 0.707 = -3$$

Again we would say that the attenuation within the line was −3 dB.

Electronic systems can have both gains and losses in the same circuit. Figure 1-21 denotes a circuit in which there is a −2dB loss, a +20 dB gain, and a −3 dB loss. The circuit has an overall gain of +15 decibels.

Communication systems often have to be designed to accommodate a combination of analog and digital signals. The designer and cabling technician are often required to deal with twisted pairs, coaxial cables, and/or fiber-optic cables. The techniques to perform these designs are discussed in Chapters 2, 3 and 4.

1.15 Insulation of Conductors

Insulation is the nonconductive material that encases a wire or cable. Insulation materials are comprised of compounds that have properties that are UL (Underwriters Laboratories) rated and CSA (Communication Society of America) approved to prevent certain environmental hazards while satisfying special electrical requirements. The electrical requirements might be fire resistance, weather resistance, pressure resistance, and so on. The environmental hazards might be that the insulator gives off toxic gases in a fire. Wires are individually insulated for ratings such as minimum breakdown voltage, wiring capacitance, maximum temperature, and so on. The primary purpose of any insulation is to prevent the short circuiting of wires to other wires or ground which could cause signal loss, damage to equipment, and possibly fire.

When more than one conductor is bundled into a cable, the insulating material for the cable is called a **jacket.** The purpose of the jacket, other than holding the wires together, is to protect the cable.

Wire and cable insulating coverings are made from several insulating materials and compounds. Insulating coverings are rated by the **Underwriters Laboratories,** a private rating company that is the industry standard for rating of consumer products for properties such as electrical characteristics, heat resistance, chemical characteristics, reliability, and safety.

The following is a list of the most common insulating materials used in the insulation of cables and their properties.

Vinyl

Vinyl is sometimes referred to as PVC or polyvinylchloride. Certain formulas have temperature ratings from $-40°$ C to $+105°$ C rating. Other common vinyls may have $-20°$ C to $+60°$C. There are many formulations for different applications. The formulation affects both the electrical properties and the pliability, which can vary from rock hard to puttylike.

Polyethylene

Has excellent electrical properties with a low dielectric value (low capacitance). Flexibility can vary from soft to rock hard. This insulation has excellent moisture resistance and can be formulated to withstand extreme outside weather conditions.

Teflon

This material has excellent electrical properties, temperature range, and chemical resistance. The material is not suited for high-voltage application or for an environment within nuclear radiation. The cost of Teflon^{TM} insulation is approximately 10 times that for comparable vinyl insulations.

Polypropylene

This insulation is similar in electrical properties, but is typically harder than polyethylene. It is suitable for thin wall insulation. Most UL ratings call for $60°$ C.

Silicone

This is a very soft insulation with a temperature range of $-80°C$ to $+90°C$. It has excellent electrical properties along with ozone resistance, low moisture absorption, weather resistance, and radiation resistance. However, it has low mechanical strength, poor scuff resistance, and cost from $5.00 to $8.00 per pound as compared to $1.00 per pound for other insulations.

Neoprene

The maximum temperature range of this material can vary from $-55°C$ to $+90°C$. The electrical properties are not as good as other insulating materials, resulting in the necessity for thicker insulation. Typically this material is used as coating for separate lead wires or cable jackets.

TABLE 1-3 Comparative Properties of Rubber Insulation

	Rubber	Neoprene	Hypalon (Chloro-sulfonated Polyethy-lene)	EPDM (Ethylene Propylene Diene Monomer)	Silicone
Oxidation resistance	F	G	E	G	E
Heat resistance	F	G	E	E	O
Oil resistance	P	G	G	F	F,G
Low temperature flexibility	G	F,G	F	G,E	O
Weather, sun resistance	F	G	E	E	O
Ozone resistance	P	G	E	E	O
Abrasion resistance	E	G,E	G	G	P
Electrical properties	E	P	G	E	O
Flame resistance	P	G	G	P	F,G
Nuclear radiation resistance	F	F,G	G	G	E
Water resistance	G	E	G,E	G,E	G,E
Acid resistance	F,G	G	E	G,E	F,G
Alkali resistance	F,G	G	E	G,E	F,G
Gasoline, kerosene, etc. (aliphatic hydrocarbons) resistance	P	G	F	P	P,F
Benzol, Toluol, etc. (aromatic hydrocarbons) resistance	P	P,F	F	F	P
Degreaser solvents (halogenated hydrocarbons) resistance	P	P	P,F	P	P,G
Alcohol resistance	G	F	G	P	G

P = poor   F = fair   G = good   E = excellent   O = outstanding

These ratings are based on average performance of general purpose compounds. Any given property can usually be improved by the use of selective compounding.

Courtesy, Belden Corporation.

TABLE 1-4 Comparative Properties of Plastic Insulation

	PVC	Low-Density Poly-ethylene	Cellular Poly-ethylene	High-Density Poly-ethylene	Poly-propy-lene	Polyure-thane	Nylon	Teflon
Oxidation resistance	E	E	E	E	E	E	E	O
Heat resistance	G,E	G	G	E	E	G	E	O
Oil resistance	F	G	G	G,E	F	E	E	O
Low temperature flexibility	P,G	G,E	E	E	P	G	G	O
Weather, sun resistance	G,E	E	E	E	E	G	E	O
Ozone resistance	E	E	E	E	E	E	E	E
Abrasion resistance	F,G	F,G	F	E	F,G	O	E	E
Electrical properties	F,G	E	E	E	E	P	P	E
Flame resistance	E	P	P	P	P	P	P	O
Nuclear radiation resistance	G	G	G	G	F	G	F,G	P
Water resistance	E	E	E	E	E	P,G	P,F	E
Acid resistance	G,E	G,E	G,E	G,E	E	F	P,F	E
Alkali resistance	G,E	G,E	G,E	G,E	E	F	E	E
Gasoline, kerosene, etc. (aliphatic hydrocarbons) resistance	P	P,F	P,F	P,F	P,F	G	G	E
Benzol, Toluol, etc., (aromatic hydrocarbons) resistance	P,F	P	P	P	P,F	P	G	E
Degreaser solvents (halogenated hydrocarbons) resistance	P,F	P	P	P	P	P	G	E
Alcohol resistance	G,E	E	E	E	E	P	P	E

P = poor   F = fair   G = good   E = excellent   O = outstanding

These ratings are based on average performance of general purpose compounds. Any given property can usually be improved by the use of selective compounding.

Courtesy, Belden Corporation.

Rubber

Both natural rubber and synthetic rubber compounds can be used for insulation and cable jackets. The material is formulated for many different applications and many different temperature ranges.

Table 1-3 presents the properties of rubber insulations.

Table 1-4 is a summary of the properties of plastic insulations.

Table 1-5 gives the nominal temperature range of various insulating materials when used as wire insulation or cable jackets.

TABLE 1-5 Nominal Temperature Range for Insulating and Jacketing Compounds

Compound	Normal Low	Normal High	Special Low	Special High
Chlorosulfonated polyethylene	−20°C	90°C	−40°C	105°C
EPDM (ethylene propylene rubber)	−55°C	105°C	—	—
Neoprene	−20°C	60°C	−55°C	90°C
Polyethylene	−60°C	80°C	—	—
Polypropylene	−40°C	105°C	—	—
Rubber	−30°C	60°C	−55°C	75°C
Teflon	−70°C	200°C	—	260°C
Vinyl	−20°C	80°C	−55°C	105°C
Silicone	−80°C	150°C	—	200°C
Halar	−70°C	150°C	—	—

Summary

The proper planning and installation of telecommunication wiring is a complex task, not to be attempted without the skills and knowledge necessary to complete the task successfully. The material in the following chapters, if studied, will greatly improve your chance of a successful installation.

Questions

1. What is the most often used unit for cable signal loss?
2. What causes cross-talk?
3. Why do wires have capacitance?
4. List several factors that would be considered in the selection of wiring insulation.
5. Define the term analog and give an example of an analog signal.
6. Define the term bit.
7. Define the term digital signals and give an example.

CHAPTER 2

Transmission Media Twisted Pair

2.1 Introduction

As a variety of data processing equipment is added to a company's inventory, management will recognize the need to tie the units together for better utilization and economy. This leads most companies to develop the equipment into some form of a **network.** When individual equipment such as host computer, PCs, workstations, and printers are connected into a common data sharing system, it is called a **local area network (LAN).** The connection of a LAN requires specialized wiring to assure the transmission of telecommunication data between the various equipments and proper operation of each unit. The wiring used is usually bundled together into cables.

A local area network is a telecommunication system within a building or a complex of buildings (campus). However, the area covered by a network or LAN is rapidly changing to include locations at great geographical distance from each other. For example, a company might tie a telecommunication system into a LAN to include business offices in New York and manufacturing facilities in Los Angeles or Korea. The on-site telephone and data communications would be via twisted-pair, coaxial, or fiber-optic cables. Information exchange between the business and manufacturing facilities could be via satellites, phone lines, or in some cases, rented fiber-optic cables. Some local telephone companies provide expansion facilities by forming entire areas into LANs on which lease time is

Figure 2-1 A twisted-pair cable.

available. Networks that cover more than local area (a campus) are called wide area networks (WANs).

In Chapters 2, 3, and 4 we will examine the wiring media used to connect a LAN in a single floor, a multistory building, or a campus. Chapter 2 focuses on copper-type *twisted pair* cables; Chapter 3 is devoted to coaxial cables; and Chapter 4 is devoted to fiber-optic cabling; followed by a summarized comparison of the three types of wiring media.

2.2 Twisted-Pair Cabling

Twisted-pair wire is the most often used cabling in the telephone industry. It is comprised of two insulated copper wires twisted approximately 20 turns per foot (Figure 2-1). The most common telephone cable uses twisted-pair wiring. Telephone cable is comprised of two solid copper wires, #24 gauge, each insulated and twisted in a pair. Usually, two sets of twisted-pair wires are enclosed in a tough gray outer insulating jacket. Each of the four wires is color coded, usually red, green, black, and yellow. The pairs are red and green, and black and yellow. Only two of the wires are needed for basic telephone functions, often leaving the other two wires for a second line or other functions. Key systems and other specialized telephone systems require four wires.

Spare twisted telephone wires can be used for many data transfer functions, and many manufacturers are producing devices that function effectively on twisted-pair wires.

Twisted-pair wiring specially designed for data cable are usually between 22 and 26 gauge as measured by the American Wire Gauge (AWG). The AWG number is the standard American Wire Gauge number which represents a host of information about each wire (Table 2-1) such as the AWG number of the wire, the diameter of wire, and the resistance and weight of the wire. Notice from the table that the larger the wire gauge number, the smaller the diameter of the wire and the higher the resistance of the wire. For example, we note from the Table 2-1 that #22 wire has a diameter of 25.36 mils (25.36 1/1000 inches) or 0.02535 inches and #24 wire has a diameter of 20.1 mils. The resistance of #22 wire is 16.14 ohms per 1000 feet and the resistance of #24 wire is 25.67 ohms per 1000 feet. Of course long runs of #22 wire would be considerably heavier than #24 wire. On the other hand, because of the higher resistance of the #24 wire, the signal would be attenuated considerably more over a #24 wire as compared to the same run of #22 wire. When "in-place" telephone cabling is used for data cabling, the installed #24 wire will suffice for many applications. However, given a choice, the cable installer should use #22 wire to reduce the effects of wiring resistance on signal transmission.

TABLE 2-1   A Partial Table of the American Wire Gauge (AWG) Showing the Characteristics of Wire

Gage No.	Diameter (mils)	Cross Section Circular (mils)	Cross Section Square inches	Ohms per 1000 Feet 25°C (= 77°F)	Ohms per 1000 Feet 65°C (= 149°F)	Pounds per 1000 Feet	Current-carrying capacity @ 700 CM per Amp	Ohms per 1,000 Ft @ 20°C
18	40.0	1620.0	0.00128	6.51	7.51	4.92	2.93	5.064
19	36.0	1290.0	0.00101	8.21	9.48	3.90	2.32	6.385
20	32.0	1020.0	0.000802	10.4	11.9	3.09	1.84	8.051
21	28.5	810.0	0.000636	13.1	15.1	2.45	1.46	10.15
22	25.3	642.0	0.000505	16.5	19.0	1.94	1.16	12.80
23	22.6	509.0	0.000400	20.8	24.0	1.54	0.918	16.14
24	20.1	404.0	0.000317	26.2	30.2	1.22	0.728	20.36
25	17.9	320.0	0.000252	33.0	38.1	0.970	0.577	25.67
26	15.9	254.0	0.000200	41.6	48.0	0.769	0.458	32.37
27	14.2	202.0	0.000158	52.5	60.6	0.610	0.363	40.81
28	12.6	160.0	0.000126	66.2	76.4	0.484	0.288	51.47

2.3 Cross-Talk on Twisted-Pair Cable

Twisted-pair cable, like any other copper cable, generates electromagnetic energy as signals pass through it, and picks up signals called cross-talk from electromagnetic waves generated from other sources. As we stated in Chapter 1, electromagnetic pickup can be from any electrical device, such as audio or data lines, motors, fluorescent light, and computers, to name a few.

The term cross-talk, as you may remember from Chapter 1, originated from the phenomenon of the conversation from an adjacent telephone line being slightly audible on another line. However, the term has been expanded to mean any type of electromagnetic interference that can cause signal interference. While electromagnetic pickup is usually only disturbing in telephone conversations, it can be devastating to digital data signals. The noise spikes caused by external electrical devices can mask some of the bits or appear as additional bits that can confuse the data coding of the computer. Figure 2-2 illustrates how the signals from one line can magnetically couple to another.

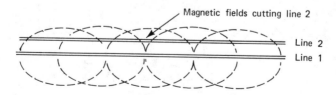

Figure 2-2   An illustration of how cross-talk can occur between transmission lines.

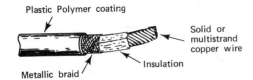

Figure 2-3 An example of shielded twisted-pair cable.

Twisted-pair cable can be shielded to reduce electromagnetic pickup and radiation by a metallic braid as shown in Figure 2-3.

2.4 Shielding of Transmission Lines

Shielding is an important part of the installation of any cabling and aids in producing a low signal-to-noise ratio in the equipment. The signal-to-noise ratio is the amount of signal voltage or power in ratio to the amount noise voltage or power. When new equipment is installed or cables or equipment is moved, the grounds to the cables and the equipment must be tested for low impedance.

The metallic shield of a cable must be grounded for the shielding to be effective in preventing electromagnetic energy from being emitted from the cable or picked up from outside sources. Shielding is most effective when each pair of wires is shielded separately. Overall shielding of a multiple conductor cable is only effective in keeping internally produced electromagnetic radiation within the cable and preventing interference from outside the cable. It does nothing to prevent cross-talk between wires within the cable.

To be most effective, shielding for telecommunication cables should be grounded at the source only. It would seem that if shielding is effective in limiting electromagnetic radiation from data communication wires and pickup from other sources, more would be better. However, grounding of the shield at both ends would create an electrical path through the ground between the source and the destination equipment, as depicted in Figure 2-4. The electrical path between equipment is called a **ground loop** and results in a difference of potential (voltage) between the source and the destination equipment. This voltage, called the

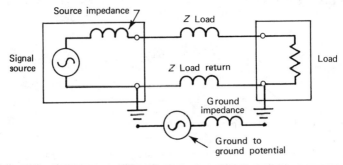

Figure 2-4 Shielding transmission lines at both ends creates a ground loop.

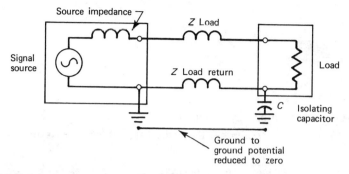

Figure 2-5   Using a capacitor to produce an effective ground at both the source and the destination.

ground loop potential, can be as much as several volts. Ground loop voltage causes a current to flow in the low resistance of the shield, resulting in electromagnetic waves that are picked up as noise by the transmission wires.

While it is desirable to shield cables at only the source, certain equipment and devices must be grounded to operate correctly. One possible solution to satisfying this need and the resulting grounding at both ends of the cable is to place a capacitor between the earth ground connection and the equipment ground (Figure 2-5). The capacitor *blocks* the dc voltage from causing a ground loop voltage while passing the varying electrical signals to ground.

Another possible solution to the multiple ground problem is the use of an isolation transformer, as shown in Figure 2-6. The isolation transformer has no physical connection between the input and the output and therefore has no return ground loop path.

Quality shielded pair cable is expensive. However, the expense is sometimes justified as such cables must meet rigid manufacturing specifications and consistencies such as

1. the diameter and the strength of each conductor
2. the properties of the insulation
3. twisting of each pair

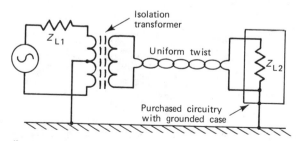

Figure 2-6   Using an isolation transformer to prevent a ground loop.

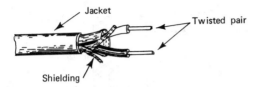

Figure 2-7   An example of IBM twin-axial shielded cable.

4. shielding of each pair by a metallic foil

5. shielding and then sealing of the entire cable by an outer layer of insulation

An example of quality shielded is the IBM twin-axial cabling for IBM cabling systems. Twin-axial cable is comprised of four wires; a red and green twisted pair, and a black and orange twisted pair; encased in a braided shield and covered by an outer insulation jacket (Figure 2-7).

In summary, shielded cable overall is subject to less cross-talk interference and radiates less electromagnetic signal. It is therefore less subject to loss of communication data that would be caused by stray pickup from other cables. Shielded cables also offer greater information security since they radiate less electromagnetic signal to other cables.

Shielding is not perfect, and the best of efforts will still result in some level of noise on a transmission line. The amplitude of the noise that occurs on a line in respect to the signal information is called the signal-to-noise ratio. Both analog and data equipment are rated as to the level of signal to noise that can be tolerated and still have the equipment perform its function.

Either shielded or unshielded twisted pair can be bundled into cables with hundreds of wire pairs as shown in Figure 2-8.

2.5 Applications and Functions of Twisted Pairs

As we noted earlier, twisted-pair cable has been used for voice transmission by telephone for years. Each pair of wires can carry more than one voice channel. This is accomplished by placing (modulating) the voice frequencies (20 to 3000 Hz) on a higher carrier frequency. A carrier frequency is a higher frequency on

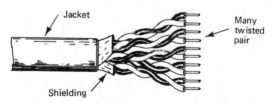

Figure 2-8   An illustration of a cable with many shielded wires.

which the intelligence or information is induced. Each carrier has a different frequency with an allotted voice bandwidth. The audio signals are modulated on to
the carrier resulting in an envelope of frequencies. Although the different audio
signals are within the same frequency range they do not interfere with each other
as the act of modulation places them within the frequency range as their carrier.

Modulation is the process of modifying a carrier frequency in rhythm to the
audio frequency. The modification of the carrier (modulation) can be by a change
of amplitude, a change of frequency, or a change of phase.

The modulation of a carrier by another signal produces other frequencies.
For example, a carrier frequency of 1 megahertz (MHz) amplitude-modulated by a
1 kilohertz (kHz) audio signal would result in the following frequencies:

$$F_{carrier}, F_{audio}, F_{carrier} + F_{audio}, F_{carrier} - F_{audio}$$
$$1 \text{ MHz}, 1 \text{ kHz}, 1.001 \text{ MHz}, \text{ and } 999 \text{ kHz}$$

These frequencies make up the modulation envelope and the bandwidth of the modulation would be 999 kHz to 1.001 MHz or 2000 Hz around the 1 MHz carrier.

The modulation of any two frequencies (F_1 and F_2) results in the development of the two new frequencies: the sum and the difference of the carrier and the
modulating frequency. Figure 2-9 illustrates the three methods of modulation.
Figure 2-9a denotes amplitude modulation where the amplitude of the carrier is
varied by the modulation. Figure 2-9b illustrates frequency modulation where the
frequency of the carrier is varied in rhythm with the modulating frequency. Figure
2-9c illustrates frequency modulation where the frequency of the carrier is varied
in rhythm with the modulating frequency. Figure 2-9c illustrates the effects of
phase modulation where the phase of the carrier is shifted in rhythm with the
modulation. Phase modulation changes the time relation of the carrier. When the
modulating signals are pulses the carrier is turned on and off (pulse modulation).

Modulation makes each frequency channel unique much as radio stations
or TV channels. This type of channeling is called **frequency division multiplexing (FDM).**

One wire pair can carry as many as 12 voice channels simultaneously using
high-frequency carrier signals. A disadvantage is that the higher the frequency of
the carrier, the greater the amount of signal attenuation. This results in the requirement of repeaters along long lines to amplify the signal.

When several signals are transmitted over the same media but at different
times the method is called **time division multiplexing (TDM).** With TDM the
signals are transmitted at different time frames that are synchronized between
the transmitter and the receiver. For example, three terminals that require access
to the same host computer appear to each be in direct connection with the computer. TDM does not change the frequency of the signals and does not create
any new signals. Figure 2-10 illustrates time division multiplexing. Sophisticated
devices that control multiplexing at each end of a telecommunication system are
called **multiplexers.**

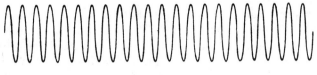

a. Carrier

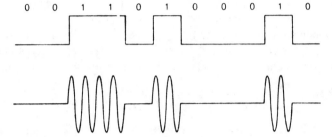

b. Amplitude Modulation

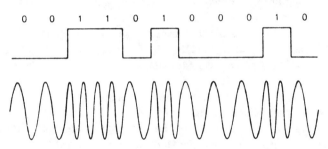

c. Frequency Modulation

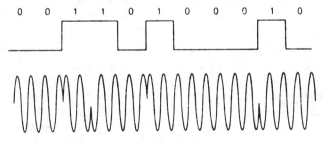

d. Phase Modulation

Figure 2-9 Four forms of modulation:
(a) amplitude modulation
(b) frequency modulation
(c) phase modulation
(d) amplitude modulation with pulses.

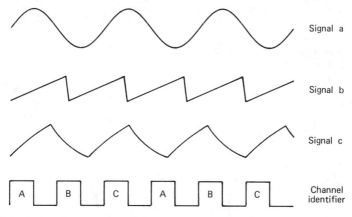

Figure 2-10   Time division multiplexing, the three signals are transmitted at different times.

2.6 Special Applications of Twisted Pairs

Special applications of twisted-wire pairs require that various modifications be made to accommodate these applications. For example, for high-electrical-noise environment shielded twisted pair is constructed with each twisted pair being shielded by metal braid or metal mylar film.

Applications of twisted-pair cables are workstation cables, distribution (feed) cables, riser (vertical distribution), and cross-connected (jumper) cable.

Inside station cable has 2 to 12 pairs of wires for 1A2 mechanical key systems. The insulation is polyvinylchloride (PVC) or plenum-rated Teflon™ for installation within the high-temperature areas of heat ducts.

Distribution cable is usually manufactured in 25 pairs and is common in 25–200 pairs. Special orders are available from 25–600 pairs. The insulation is usually gray PVC. However, Teflon™-coated pairs can be ordered for resistance to high temperature.

Table 2-2 gives a summary of two-wire cables: voice and data cables and twin axial. Cable can be ordered in 200 to 3000 pairs.

Direct-bury cable (jelly-filled shielded) comes in 2 to 3600 pairs. Armored cable has a thick, strong outer jacket to protect the cable pairs.

The insulation on all these cables is available in various colors for identification.

2.7 Twisted-Pair Cable Termination

Regardless of the type of twisted-pair cable that is used, each wire must be terminated at the source, at the destination, and at splices in long cables. Cables can be purchased with a variety of connectors for almost any application. These same

TABLE 2-2 Twinaxial Cables

Description	MONCO CBL	UL Style	Conductor Material Stranding Diam (Inch)	Dielectric Material Diameter (Inch)	Shield Material % Coverage	Jacket Material Nom Diam (Inch)	Nom Vel of Prop.	Nom Imped (ohms)	Nom Capac (pf/ft)	Nominal Attenuation MHz / dB/100'
Appletalk[1] Non-plenum	6242	2726	22 AWG 7/30 T.C.	PE .056	T.C. Braid 85% Coverage	PVC Snow Beige .183	66%	78	20.7	NA NA
Appletalk[1] Plenum	6228	NEC 725-38 (b) (3)	22 AWG 7/30 T.C.	FEP³ .057	T.C. Braid 85% Coverage	FEP³ Beige .184	70%	78	18.9	NA NA
78 Ohm Non-plenum CIM	6453	2092	20 AWG 7/28 T.C.	PE	T.C. Braid 93% Coverage	PVC Blue .242	65%	78	20.1	1→.6, 10→2.1, 20→3.0, 50→5.0, 100→7.5, 200→11.0, 400→16.0
78 Ohm Plenum	6292	NEC 725-38 (b) (3)	20 AWG 7/28 T.C.	FEP³	T.C. Braid 93% Coverage	Flouropolymer Blue Tint .220	70%	78	19.5	
78 Ohm RG 108-Type Factory Data Bus	6502	2582	20 AWG 7/28 T.C.	PE	Alum. Tape w/Drain + T.C. Braid 57% Coverage	PVC Blue .243	65%	78	20.1	1→.6, 10→2.1, 20→3.0, 50→5.0, 100→7.5, 200→11.0, 400→16.0
RG 22 B/U-Type	6503	2092	18 AWG 7/.0152 1 B.C. + 1 T.C.	PE .285	2 T.C. Braids 96% Coverage	PVC Black .420	65%	95	16.5	1→.3, 10→.9, 20→1.3, 50→2.1, 100→3.0, 200→4.5, 400→6.3

Description	Part No.		Conductor	Insulation	Shield	Jacket	Coverage	Impedance	Capacitance	Freq.	Atten.
100 Ohm IBM System 36[2] Non-plenum	6504	2498	20 AWG 7/28 1 B.C. + 1 T.C.	PE .240	T.C. Braid 95% Coverage	PVC Black .330	65%	100	15.6	1 10 20 50 100 200 400	.4 1.1 1.5 2.5 4.1 6.4 10.2
100 Ohm IBM System 36[2] Plenum	6505	NEC 725-38 (b) (3)	20 AWG 7/28 1 B.C. + 1. T.C	FEP[3]	T.C. Braid 95% Coverage	Fluoropolymer Black Tint .282	66%	100	15.4		
124 Ohm Programmable Controller Type	5864	2092	25 AWG 7/33 T.C.	PE	Alum. Tape + T.C. Drain	PVC Blue .240	65%	124	12.6	1 10 20 50 100 200 400	.6 1.7 2.3 3.6 5.0 6.9 9.6
150 Ohm Factory Communication	6508	2668	22 AWG 19/34 T.C.	Foam Polypro	Alum. Tape + T.C. Drain	PVC Black .350	75%	150	9.0	1 10 20 50 100 200 400	.3 1.3 2.3 3.0 4.3 6.2 8.8
124 Ohm Factory Communication	6506	2448	16 AWG Solid Bare Copper	Foam PE	Alum. Tape + T.C. Braid 95% Coverage	PVC Black .440	75%	124	10.9	1 10 20 50 100 200	.27 .89 1.3 2.0 2.9 4.0

TABLE 2-2 (*Continued*)
Voice and Data Cables

Description	MONCO CBL	UL Style	Conductor Material Stranding Diam (inch)	Dielectric Material Diameter (Inch)	Shield Material % Coverage	Jacket Material Nom Diam (inch)	Nom Imped (ohms)	Nominal Attenuation MHz	Nominal Attenuation dB/100'
TYPE I 2 Data Pairs Non-Plenum	6100-748C	Subject 13	22 AWG Solid Bare Copper	Foam PE .090	Alum. Foil Tape each Pair, plus overall, T.C. Braid 65% Coverage	PVC Black .380	150/4MHz	4 16	.66 1.37
TYPE I 2 Data Pairs Plenum	6100-749C	NEC 725-38 (b) (3)	22 AWG Solid Bare Copper	Foam FEP .100	Alum. Foil Tape each Pair, plus Overall T.C. Braid 65% Coverage	Fluoropolymer Black Tint .365	150/4MHz	4 16	.66 1.37
TYPE II 2 Data Pairs 4 Voice Pairs Non-Plenum	6100-739C	Subject 13	22 AWG Solid Bare Copper	2-Pair Foam FRPE .100 4-Pair Solid FRPE	2-Pair w/ individual Alum. Tape, plus overall T.C. Braid 65% Coverage	PVC Black .495	Data Pairs: 150/4MHz Voice Pairs: 100/4MHz	Data Pairs: 4 16 Voice Pairs: 1kHz 772/kHz	.66 1.37 .04 .51
TYPE II 2 Data Pairs 4 Voice Pairs Plenum	6100-738C	NEC 725-38 (b) (3)	22 AWG Solid Bare Copper	2-Pair Foam FRPE .100 4-Pair Solid FEP .045	2-Pair w/ individual Alum. Tape, plus overall T.C. Braid 65% Coverage	Fluoropolymer Black Tint .420	Data Pairs: 150/4MHz Voice Pairs: 100/4MHz	Data Pairs: 4 16 Voice Pairs: 1kHz 772/kHz	.66 1.37 .04 .51
TYPE 6 2-Pair Office Grade Data Cable	6100-743C	Subject 13	26 AWG 7/34 Bare Copper	Foam PE .070	Overall Alum. Tape plus T.C. Braid 65% Coverage	PVC Striated Black .327	150/4MHz	4 16	1.0 2.01

1 = Apple Computer trademark   2 = IBM Corp. trademark   3 = FEP—registered trademark of E.I. DuPont Co. Note: Consult Graybar for twinaxial variations not shown above

Twisted-Pair Data Cables

Description	MONCO CBL	UL Style	Number of Pairs	Conductor Material Stranding Diam (Inch)	Dielectric Material Diameter (inch)	Colors	Jacket Material Non Diam (Inch)	Shield Material % Coverage	Nom Imped (ohms)
Usernet Pair Non-Plenum	6734	2509	1	20 AWG 10/30 T.C.	PVC .072	Black + Red	PVC Gray .190	NA	NA
Usernet Pair Non-Plenum Shielded	5624	2092	1	20 AWG 7/28 T.C.	PE .068	Black + Clear	PVC Gray .204	Alum. Tape plus 20 AWG Drain	
Usernet Pair Plenum	6507	NEC 725-38 (b) (3)	1	20 AWG 7/28 T.C.	Tefzel[2] .066	Black + Red	Tefzel[2] Clear .172	NA	NA
Omninet[3] Pair Non-Plenum	6408	2919	1	24 AWG 7/32 T.C.	Foam Polypropylene .064	Black + Red	PVC Gray .201	NA	135
IEEE802.3 1-Base-5	6513	20245	4	24 AWG 7/32 T.C.	Semi-rigid PVC .039	Telephone Standard	PVC White .205	NA	95
IEEE802.3 1-Base-5 Plenum	0586-03	NEC 725-38 (b) (3)	4	24 AWG 7/32 T.C.	FEP .0075	Telephone Standard	Kynar	NA	95
Type 3 Token Ring	6512	NA	4	24 AWG Solid Bare	Semi-rigid PVC .033	Telephone Standard	PVC Gray .170	NA	100

1 = Trademark of Sperry Corp.—Usernet 2 = Trademark of E.I. DuPont Co.—Tefzel 3 = Trademark of Corvus Systems—Omninet
Note: Please consult Graybar for designs not shown here.

connectors can be purchased and installed on cables. The following is a sampling of cable connectors:

- 2–12 pair, with 3 to 6 pair the most common
- Duplex jacks (two jacks) usually have voice and data combination
- Surface-mount connectors
- Flush-mount connectors
- RJ11 with six positions and four contacts
- RJ45 with six positions and six contacts
- RJ45 with eight positions and eight contacts

Multifunction custom-built connectors and terminals are available in the push-on and screw type. Some equipment manufacturers such as IBM, AT&T, and Digital Equipment require custom connectors. A great variety of termination connectors are available.

Grounding of the cable shield is an important part of the termination of shielded twisted-pair cable. Special grounding techniques were covered in Chapter 1. However, there are techniques that apply specifically to twisted pair, for example, the continuation of a ground through a pair of punch-down blocks (Figure 2-11). The ground from each cable is connected to a twisted pair of wires that completes the connection. The twist in the connection prevents any electromagnetic pickup in the ground extension between the blocks.

When a cable enters or exits a building, a lightning arrestor must be installed between the lines and ground. The device acts as an open circuit unless a very large voltage, such as a lightning bolt, surges between the wires and ground. A large voltage surge causes a short circuit to ground, which gives a low-resistance path across which the energy can dissipate. Figure 2-12 depicts one such type of lightning protector.

2.8 Distribution Frames

When many cables are installed for a system they are usually run from a **main distribution frame (MDF)**. The MDF is a point at which the cables can be added to, deleted from, or rerouted in the system. The MDF (Figure 2-13) is usually located in a room or area called a closet. A large system may have auxiliary or **intermediate distribution frames (IDF)**.

A typical AT&T MDF has 66 blocks vertically mounted on 89 B brackets. These frames can be purchased individually or per standard manufactured sizes, with 25 pairs and 50 pair-split blocks being the most common. An example of a telephone termination block is shown in Figure 2-14.

Special application blocks are available. For example, AT&T 110 PDS equipment is horizontally orientated, high density with a complete line of blocks,

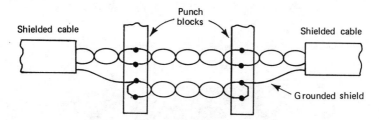

Figure 2-11   An example of a ground connection for twisted pair between punch-down blocks.

mounting equipment, and labeling. Auxiliary equipment such as patch cords, racks, and patching tools are also available in this line.

Another line of MFD or IFD equipment is BIX blocks. This line is horizontally oriented, and high density. A complete line of products such as blocks, mounting brackets, labels, patch cords, and so on, are available. Special termination tools are necessary for the installation of the cables. These will be discussed later in this chapter.

2.9 Existing Cable Systems and Compatibility

Most businesses have an existing telephone cabling system in place, which can be formed into a LAN with some data processing equipment. Some of the questions that a telecommunication manager must ask before deciding to use in-place teleco wiring are

1. Can the existing system be incorporated into the new plans?
2. Is the old equipment compatible with the new system?
3. Is current staff capable of installing new lines and new equipment and testing the system?
4. Should the LAN use a telephone system, a PBX, a key system, and so on?
5. Should the computer/data equipment use workstation wiring (moduLAN, RS-232, etc.)?
6. Should the old twisted pair be connected by coaxial conversion via balun to coax?
7. Should patch panels be used?

Figure 2-12   An example of a lightning arrestor that would be used on a twisted-pair line. Courtesy, Anister Brothers, Inc.

Figure 2-13 A main distribution frame. Courtesy, Nevada Western Corp.

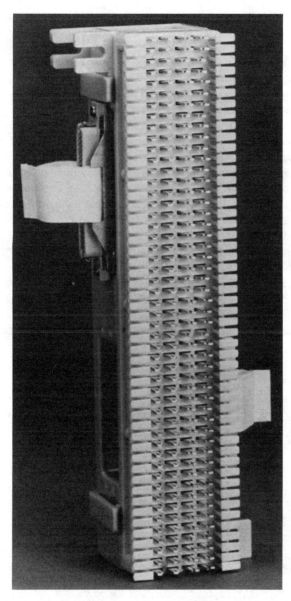

Figure 2-14   An example of a telephone termination block. Courtesy, Nevada
Western Corp.

2.10 Electrical Characteristics of Twisted-Pair Cables

The electrical characteristics of cables were discussed in general in Chapter 1. Here we will discuss those particular to twisted-pair cables. Table 2-3 is a chart of the continuous current rating of conductors at four different temperatures for different sizes of wire. For example, a #24 gauge wire has a rating of approximately 1.8 amperes at temperatures between 60°C and 80°C.

Table 2-4 depicts the attentuation in decibels per 100 feet for three different types of twisted-wire packaging. The important consideration here is to observe the fact that the decibel loss increases significantly as frequency of the signal on the cable increases. Note that a 10 dB power loss reduces signal power to one-tenth its original value and voltage to approximately one-third its original value.

Table 2-5b is a chart of the rise time of transmitted pulses versus the transmission distance. You may remember from Chapter 1 that capacitance of a cable is directly proportional to the length of the cable and the more capacitance a cable has results in greater rise time of the digital pulses transmitted along the cable.

Table 2-5c is a chart of the allowable bit rate of data pulses versus the transmitted distance of the signal. From the table, we see that a transmission distance of 1000 feet would allow a bit rate of 500,000 bits per second. The maximum bit rate of a cable is dependent to a great degree on the resistance of the wires and the capacitance of the cable. Note that the ratings given in the charts are for transmission distance and not the length of the cable, which would be over twice the transmission distance. When calculating the transmission distance the

TABLE 2-3   A Chart of Current vs Conductor Size for
Different Temperatures

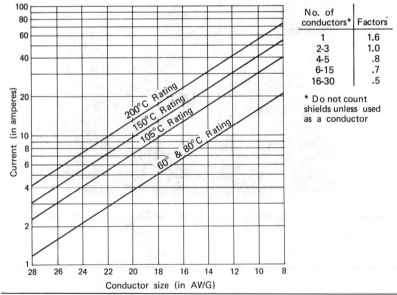

No. of conductors*	Factors
1	1.6
2-3	1.0
4-5	.8
6-15	.7
16-30	.5

* Do not count shields unless used as a conductor

TABLE 2-4   A Graph of Frequency vs Attenuation in dB for
Different Sizes of Wire

Courtesy, Belden Corp.

length of all the jumper wires connecting data processing equipment to the cable must be added to the distance.

Twinaxial transmission cable is special twisted pair cabling in which each pair of wires is individually shielded. The individual pairs are formed into a cable and the entire package is then shielded. This type of cable is more expensive than simple twisted pair. However, it offers low-loss signal transmission free of outside signal or noise electromagnetic fields. A comparison of the characteristics of twinaxial cable in Table 2-6 to that of twisted pair in Table 2-5 shows that twinaxial cable is a superior cable type.

Another consideration when making connection to twisted-pair cabling is the characteristic impedance of the cable. This is especially true when coaxial cabling is connected to twisted pair, but can be a problem when data processing equipment has an output impedance for matching to a coaxial cable. In either case the twisted pair must be matched with a device called a balun. A **balun** is a transformer device for connecting cables to other cables or devices that have a different characteristic impedance. The transformer in the balun can electrically isolate the

TABLE 2-5 Characteristics of Twisted-pair Cable: a. dB vs Frequency, b. Pulse
Rise Time vs Transmission Distance, c. Bit Rate vs Transmission Distance.

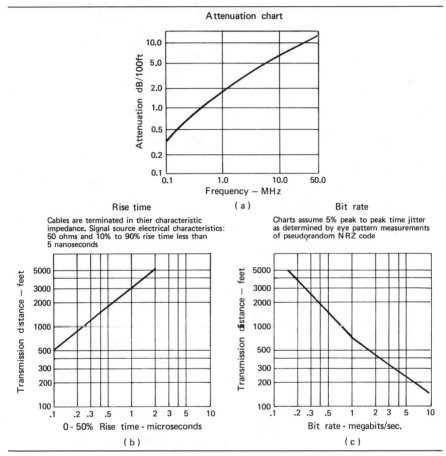

Attenuation chart

Rise time   (a)   Bit rate

Cables are terminated in thier characteristic
impedance. Signal source electrical characteristics:
50 ohms and 10% to 90% rise time less than
5 nanoseconds

Charts assume 5% peak to peak time jitter
as determined by eye pattern measurements
of pseudorandom NRZ code

(b)   (c)

Courtesy, Belden Corporation.

lines and make each appear to be matched to the correct impedance. Impedance
matching is important because a mismatch can result in serious signal loss.

To prevent cross talk, shielding should also be extended to the connectors as
they can be a source of electromagnetic pickup. The effectiveness of the shielding
for both cables and connectors depends upon the frequency of the electromagnetic
interference. Table 2-7 depicts the effects of frequency on the shielding ability of
an RS-232 type connector. Notice that shielding is generally less effective for all
types of materials as the frequency of the outside interference increases. For example, in the case of copper foil shielding, there is a drop of almost 10 decibels
between 60 MHz and 50MHz. This means that the shielding is only one-eighth as
effective at 50MHz as at 60MHz.

TABLE 2-6 Characteristics of Twinaxial Cable: a. Attentuation vs Frequency,
b. Bit Rate vs Transmission Distance, and c. Frequency vs Transfer Impedance.

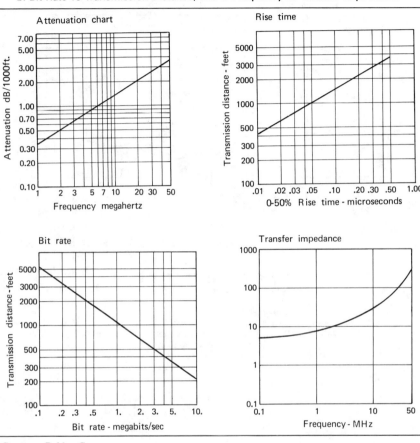

Courtesy, Belden Corp.

2.11 Tools for the Installation of Twisted-Pair Cables

The installation of each type of cabling requires special tools. Here we will discuss some of those required for the installation of twisted-pair cables. The diagonal pliers depicted in Figure 2-15 are typical of electrical wire cutters. Diagonal pliers are available in many different sizes to handle various size wires. The electrical pliers shown in Figure 2-16 can be used for cutting wires and as a pliers to hold wires or parts in place.

The electrical wire stripper-cutter crimping tool shown in Figure 2-17 can serve several functions:

1. The tips can be used as wire cutters.

2. The inside of the handles can be used as wire strippers.

TABLE 2-7 A Comparison of Shielding Effectiveness vs Frequency for a Shielded Connector

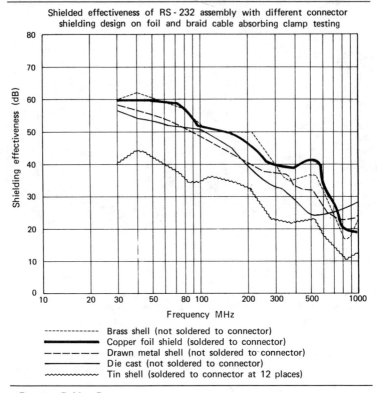

Shielded effectiveness of RS-232 assembly with different connector shielding design on foil and braid cable absorbing clamp testing

Shielding effectiveness (dB)

Frequency MHz

```
------------   Brass shell (not soldered to connector)
━━━━━━━━━   Copper foil shield (soldered to connector)
— — — —   Drawn metal shell (not soldered to connector)
───────   Die cast (not soldered to connector)
∿∿∿∿∿∿   Tin shell (soldered to connector at 12 places)
```

Courtesy, Belden Corp.

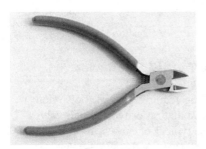

Figure 2-15 Twisted-pair installation tools

Figure 2-16 An example of electrical pliers.

Figure 2-17   An example of wire stripper/crimper. Courtesy, Kline Tools, Inc.

3. The holes in the handles are threaded so that screws can be cut to length without ruining the threads.

4. The blades are designed for connector crimping.

The tool shown in Figure 2-18 is a punch-down tool. It is designed to punch down the copper wire of twisted-pair cables correctly in a **punch-down block.**

An example of tools that are made especially for installing wires is the combination stripper wire cutter shown in Figure 2-19.

Figure 2-20 is a hand tool for stripping the outer and inner insulation from a multiple wire cable. The outer cable is first stripped from the cable and each wire is then stripped. The connector pins then inserted over each wire and crimped with a crimping tool. Finally, the pins are pressed into the connector.

Figure 2-18   An example of a punch-down tool. Courtesy, Amp. Inc.

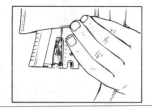

Automatic adjustment to cable area

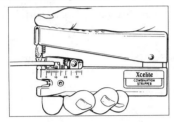

Built-in cutter

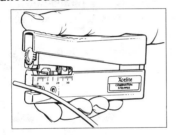

Figure 2-19 An example of a wire cutter/stripper designed for the wiring technician. Courtesy, Xcelite Corp.

There are many tools being designed to make the installation of cables easier. It is important to keep abreast of the latest in the catalogs from the vendors listed in the appendix.

2.12 Advantages of Twisted-Pair Cabling

Some of the advantages of twisted pairs over other types of cabling are the following:

1. Many developments of new applications utilize twisted pair.
2. Flexibility in that existing twisted-pair wiring system can be utilized with new applications.
3. Distribution cable can be used and all home runs will usually not be necessary.
4. Allows for simultaneous voice/data cabling and saves material and labor and therefore results in less cost.

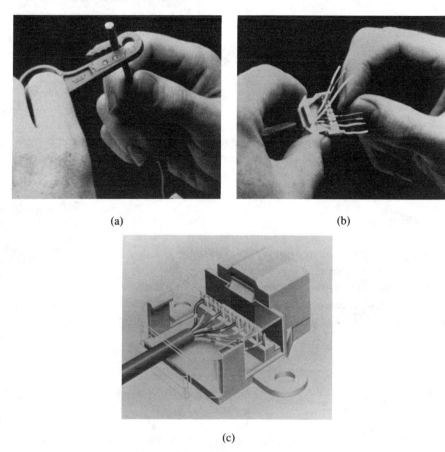

(a)   (b)

(c)

Figure 2-20   A tool for stripping outer and inner insulation from a multiple
strand cable:
a. stripping the outer insulation
b. inserting wires in connectors
c. finished connector. Courtesy, Nevada Western Corp.

5. Technicians familiar with telephone cabling can easily train for other wiring
   applications.
6. Twisted-pair wiring is often in place.

2.13 Disadvantages of Twisted-Pair Cabling

Disadvantages of twisted pair are

1. limitation in data bandwidth speed and data speed

2. possible cross-talk interference

3. less security than coax or fiber cable

Summary

Twisted-pair cables are the most common transmission media in voice and data transmission. New developments are increasing data speed and bandwidth capabilities of twisted-pair cables. Twisted-pair cables are the easiest and least expensive to install.

Questions

1. What are the limitations of twisted-pair cables?

2. What are the advantages of twisted-pair cables over fiber and coaxial cables?

3. When should twisted-pair cable be utilized for a LAN connection?

4. What is the difference between #24 and #22 twisted-pair wire?

5. What is the purpose of grounding the shield on a cable?

CHAPTER 3

Coaxial Cable

3.1 Introduction

A coaxial cable (commonly called coax) is comprised of a center conductor and an outer shielding conductor (Figure 3-1). The center conductor can be a single wire or stranded wire. Single conductors have less resistance than stranded conductors but are less flexible and can kink unless care is used in the installation. Some coaxial cables are constructed with a strain relief in the form of a small diameter fiber or plastic line wrapped around the center conductor. Coax that are used as patch panels or service bundles should have this strain relief.

The center conductor of a coax is surrounded by an insulating material, called a **dielectric,** which is in turn surrounded by an outer metallic shield. The magnetic shield serves as the return path for the electrical energy and as a shield

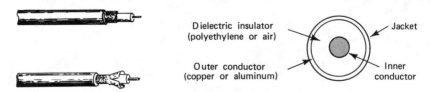

Figure 3-1 Construction details of coaxial cable:
a. an older type of braided shield
b. a TNC video type with a solid shield.

Figure 3-2 A coaxial cable with two inner cables, packaged to form two coaxial cables.

for the center conductor against electromagnetic pickup and cross-talk and to limit radiation of the signal from the center conductor. The metallic shield is covered by a tough outer insulating material. When a very large number of signals are to be transported over the same route, but over individual coaxial cables, coax can be bundled into a multicable jacket for ease of installation.

3.2 Characteristics and Construction

Coaxial cable is also constructed with two center conductors, each a coaxial cable in itself. The two coaxs' are insulated from each other, surrounded by a dielectric, enclosed by a magnetic shield and surrounded by an outer insulator (Figure 3-2). This packaging forms two coaxial cables in a small diameter.

There are several types of coaxial cables, each with the same general construction but with somewhat different electrical characteristics. Figure 3-1 denoted the two basic types of construction:

1. Solid metallic shield
2. Braided metallic shield

Table 3-1 summarizes the types of coaxial cable and their respective characteristic impedance.

Ethernet trunk cable is usually of the 50 ohm impedance type such as RG 8 or RG 58. Video cable is usually of the 75 ohm type such as RG 59. Coaxial cable is known for its wide bandwidth, less electromagnetic radiation, less cross-talk, and greater security than twisted pair. Ethernet is a local area network developed by Xerox and then sponsored by Digital Equipment Corp. (DEC) and Intel Corp. so as to become a standard in the industry. In an Ethernet system the receive and send signals are all transported along a single coaxial cable (Figure 3-3a) and coupled to various data communication equipment by a Ethernet transceiver called a **drop point** (Figure 3-3b). The cable length maximum is approximately 500 meters. The cable between the drop point and the workstation uses a

TABLE 3-1 Standard Coaxial Types and Their Characteristic Impedance

RG 6	75 Ohms 18 AWG center conductor
RG 8	50 Ohms 18 AWG center conductor
RG 11	75 Ohms 14 AWG center conductor
RG 58	50 Ohms 14 AWG center conductor
RG 59	75 Ohms 22 AWG center conductor
RG 62	92 Ohms 22 AWG center conductor

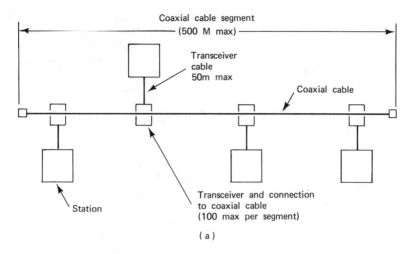

(a)

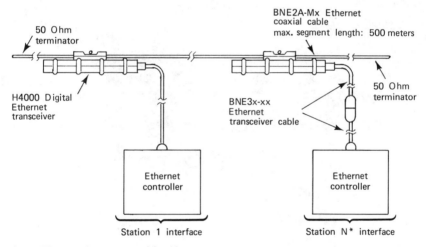

(b)

Figure 3-3 An Ethernet network system:
a. basic system
b. transceiver drop connection.

smaller coaxial cable. Ethernet coaxial cable is available in various lengths which can be connected together with barrel connectors (Figure 3-4). The barrel connector causes very little signal attenuation.

   Although all coaxial cable has basically the same construction, there are a number of types manufactured to withstand specific environmental conditions.

Figure 3-4   An example of a barrel-type connector and terminators used to connect coaxial cables. Courtesy, AMP Corp.

The size of the inner conductor, the insulation dielectric material, and the outer shield material used results in cables having different characteristic impedance. Table 3-2 is a summary of the types of coaxial cables.

We note from Table 3-2 that the characteristic impedance for coax ranges between 50 and 95 ohms. Cables are available that contain twisted pair, twinaxial, and/or coax. Table 3-3 is a summary of one manufacturer's inventory.

TABLE 3-2   A Summary of the Types of Coaxial Cables

- **Cable RG 59/U-1 (105 482 624)** is a 75 ohm coaxial cable with a 22 AWG (7x30) center conductor, a foamed polyethylene dielectric, a bare copper braid (min. 95% coverage) outer conductor and a PVC jacket. (Similar to RG 59/U type.) UL style 1354.
- **Cable RG 59/U-1A (105 521 561)** is a 75 ohm coaxial cable with a 22 AWG (7x30) center conductor, a foamed polyethylene dielectric, a bare copper braid (min. 95% coverage) outer conductor and a PVC jacket. (Similar to RG 59/U type.) UL style 1354, UL Listed Type CL2.
- **Cable RG 59/U-2 (105 482 632)** is a 75 ohm coaxial cable with a 22 AWG copper covered steel center conductor, a polyethylene dielectric, a bare copper braid (min. 80% coverage) outer conductor and a PVC jacket. (Similar to RG 59/U type commercial.) UL style 1354.
- **Cable RG 59/U-2A (105 521 579)** is a 75 ohm coaxial cable with a 22 AWG copper covered steel center conductor, a polyethylene dielectric, a bare copper braid (min. 80% coverage) outer conductor and a PVC jacket. (Similar to RG 59/U type commercial.) UL style 1354. UL Listed Type CL2 per 1987 NEC.
- **Cable RG 59/U-5 (105 482 665)** is a 75 ohm plenum coaxial cable with a 22 AWG copper covered steel center conductor, an FEP dielectric, a bare copper braid (min. 95% coverage) outer conductor and an FEP jacket. (Similar to RG-59/U type.) UL Listed Type CL2P per 1987 NEC.
- **Cable RG 62 A/U-1 (105 482 723)** is a 93 ohm coaxial cable with a 22 AWG copper covered steel center conductor, an air dielectric polyethylene dielectric, a bare copper braid (min. 95% coverage) outer conductor and a PVC jacket. (Similar to RG 62 A/U type.)
- **Cable RG 62 A/U-1A (105 521 660)** is a 93 ohm coaxial cable with a 22 AWG copper covered steel center conductor, an air dielectric polyethylene dielectric, a bare copper braid (min. 95% coverage) outer conductor and a PVC jacket. (Similar to RG 62 A/U type.) UL Listed Type CL2 per 1987 NEC.
- **Cable Ethernet™-1 (105 482 798)** is a 50 ohm coaxial cable with a 0.0855 AWG solid tinned copper center conductor, a foamed polyethylene dielectric, a foil shield bonded to dielectric, a tinned copper braid (min. 92% coverage) a foil shield, a tinned copper braid (min. 92% coverage) and a PVC jacket. (Similar to Ethernet Type.) UL style 1478 DEC approved. "DEC NET." Xerox specifications/IEEE 803.
- **Cable Ethernet-1A (105 538 037)** is a 50 ohm coaxial cable with a 0.0855 AWG solid tinned copper center conductor, a foamed polyethylene dielectric, a foil shield bonded to dielectric, a tinned copper braid (min. 93% coverage) a foil shield, a tinned copper braid (min. 90% coverage) and a yellow PVC jacket. (Similar to Ethernet Type.) UL style 1478. UL Listed Type CL2 per 1987 NEC. DEC approved. "DEC NET." Xerox specifications/IEEE 803.
- **Cable Ethernet-2 (105 482 806)** is a 50 ohm plenum coaxial cable with a 0.0855 AWG solid tinned copper center conductor, a foamed FEP dielectric, a foil shield, a tinned copper braid (min. 93% coverage), a foil shield, a tinned copper braid (min. 90% coverage) and an FEP jacket. (Similar to Ethernet.) UL Listed Type CL2P per 1987 NEC. DEC approved. Xerox specifications/IEEE 803.

3.3 Coaxial Cable Connectors and Terminations

Whenever possible, coaxial cable should be purchased with the connectors attached. However, this is seldom possible for long cable runs and points of cable drops where outputs are taken for data processing equipment. These connections will require that cable adapters be installed, and in some cases the connectors must be installed by the wiring technician.

TABLE 3-3 A Summary of Composite Building Cables

Description	MONCO CBL	UL Style	Conductor Material Stranding Diam (Inch)	Dielectric Material Diameter (Inch)	Shield Material % Coverage	Jacket Material Nom Diam Inch	Nom Vel. of Prop.	Nom Imped (ohms)	Nom Capac (pf/ft)	Nominal Attenuation (MHz / dB/100')
RG 59/U type plenum	6460	NEC 725-38 (b) (3)	.032 Copper Covered Steel	Foam FEP¹ .146	Bare Copper Braid 95% Coverage	Fluoropolymer Black .205	81%	75	16.0	50/1.8, 100/2.6, 200/3.8, 500/6.2
RG 59/U type nonplenum	6489	1354	.025 Bare Copper Covered Steel	PE .146	Bare Copper Braid 95% Coverage	PVC Black .242	65.5%	73	21.3	50/2.4, 100/3.4, 200/4.9, 400/7.1, 700/9.5, 900/10.9, 1000/12.0
RG 59/U type plenum	6490	NEC 725-38 (b) (3) 1354	.025 Bare Copper Covered Steel	FEP¹ .135	Bare Copper Braid 95% Coverage	Fluoropolymer Black Tint .206	69%	75	19.6	100/3.4, 200/4.9, 400/7.1
RG 59/U type nonplenum cellular	6234	1354	22 AWG 7/30 Bare Copper	Foam PE .146	Bare Copper Braid 95% Coverage	PVC Black .242	75%	75	18	50/2.1, 100/3.0, 200/4.5, 400/6.6, 700/8.9, 900/10.1, 1000/10.9
RG 59/U type plenum cellular	6492	NEC 725-38 (b) (3)	22AWG 7/30 Bare Copper	Foam FEP¹ .146	Bare Copper Braid 95% Coverage	Fluoropolymer Black Tint .218	81%	75	16.7	50/2.1, 100/3.0, 200/4.5, 400/6.6, 900/10.1
Dual RG 59/U type nonplenum	6491	20063	.023 Bare Copper Covered Steel	PE .146	Bare Copper Braid 95% Coverage	PVC Black .238 × .478	65.5%	75	20.7	100/3.4, 200/5.1, 400/7.5, 700/11.4, 900/12.0, 1000/12.7
Dual RG 59/U type plenum	6454	NEC 725-38 (b) (3)	.023 Bare Copper Covered Steel	FEP¹ .134	Bare Copper Braid 95% Coverage	Fluoropolymer Clear .236 × .442	69%	75	19.6	100/3.5, 200/5.1, 400/7.5, 700/11.4, 900/12.0, 1000/12.7

Cable	Cat. No.	NEC/UL	Center Conductor	Dielectric	Shield	Jacket	Coverage	Imp.	Cap.	Freq (MHz) / Atten.
nonplenum triax			Copper Covered Steel	.143	Braid 96% Coverage w/ PE insulation between braids	Black .315				100 — 2.6; 200 — 3.8; 300 — 4.8; 400 — 5.6; 900 — 8.4
RG 59/U type plenum triax	6494	NEC 725-38 (b) (3)	.032 Bare Copper Covered Steel	Foam FEP¹ .140	2 Bare Copper Braid 96% Coverage w/ Fluoropolymer insulation between braids	Fluoropolymer Black Tint .262	81%	75	16.7	50 — 1.8; 100 — 2.5; 200 — 3.6; 500 — 6.0; 900 — 8.6
RG 62 A/U type nonplenum	5770	1478	22 AWG Solid Bare Copper Covered Steel	Semi-solid PE .146	Bare Copper Braid 90% Coverage	PVC Black .242	80%	93	13.7	400 — 8.0
RG 62 A/U type plenum cellular	5727-1	NEC 725-38 (b) (3)	22 AWG Solid Bare Copper Covered Steel	Foam FEP¹ .146	Bare Copper Braid 90% Coverage	Fluoropolymer White Tint .225	80%	93	14.5	400 — 8.0
RG 8/U type nonplenum	6495	1354	11 AWG 7/19 Bare Copper	Foam PE .285	Bare Copper Braid 97% Coverage	PVC Black .405	78%	50	26.1	50 — 1.2; 100 — 1.8; 200 — 2.7; 400 — 4.2; 700 — 5.8; 900 — 6.7; 4000 — 18.0
RG 8/U type plenum	6496	NEC 725-38 (b) (3)	11 AWG 7/19 Bare Copper	Foam FEP¹ .285	Bare Copper Braid 97% Coverage	Fluoropolymer Black Tint .365	81%	50	25.1	50 — 1.2; 100 — 1.8; 200 — 2.7; 400 — 4.2; 700 — 5.8; 900 — 6.7; 4000 — 18.0
RG 11/U type nonplenum	6497	1354	.064 Bare Copper	Foam PE .285	Bare Copper Braid 95% Coverage	PE .405	78%	75	17.4	50 — 1.0; 100 — 1.5; 200 — 2.2; 500 — 3.7; 900 — 5.2

¹FEP = Registered trademark of E. I. Du Pont de Nemours & Co.

Note: Please consult Graybar for coaxial variations not shown above.

TABLE 3-3 (Continued)

Description	MONCO CBL	UL Style	Conductor Material Stranding Diam (Inch)	Dielectric Material Diameter Inch	Shield Material % Coverage	Jacket Material Nom Diam Inch	Nom Vel of Prop.	Nom Imped (ohms)	Nom Capac (pf/ft)	Nominal Attenuation MHz	dB/100'
RG 11/U type nonplenum	6498	1354	.064 Bare Copper	Foam PE .285	Alum/Tape + T.C. Braid 61% Coverage	PVC .405	78%	75	17.4	50 100 200 500 900	1.0 1.5 2.2 3.7 5.2
RG 11/U type plenum	6499	NEC 725-38 (b) (3)	.064 Bare Copper	Foam FEP[1] .285	Alum/Tape + T.C. Braid 61% Coverage	Fluoropolymer .363	80%	75	16.9	50 100 200 500 900	1.0 1.5 2.2 3.7 5.2
RG 11/U type nonplenum triax	6500	1354	.064 Bare Copper	Foam PE .285	2 Bare Copper Braid 96% Coverage w/ PE insulation between braids	PE .475	78%	75	17.4	50 100 200 300 400 900	1.0 1.5 2.2 2.8 3.3 5.2
RG 11/U type plenum quad	6501	NEC 725-38 (b) (3)	.064 Bare Copper Covered Steel	Foam FEP[1] .280	2 Alum/Tapes + 2 T.C. Braids	Fluoropolymer .389	82%	75	16.5	400	3.0
Ethernet[2] nonplenum	5688	1478 60°C	.0855 Bare Copper	Foam PE .247	Alum/Tape + T.C. Braid 90% Min. Cov. + Alum/Tape + T.C. Braid 90% Min. Cov.	PVC Yellow .405	78%	50	26.0	5 10	.38 .53

[1]FEP = Registered trademark of E. I. Du Pont de Nemours & Co.

56

Description	Part No.	NEC Rating	Conductor	Dielectric	Shield	Jacket						
Ethernet[2] plenum	5713-1	NEC 725-38 (b) (3) 125°C	.0855 Bare Copper	Foam FEP[1] .245	Alum/Tape + T.C. Braid 90% Min. Cov. +Alum/Tape +T.C. Braid 90% Min. Cov.	Kynar Flex[3] Orange .375	78%	50	26.0	5	10	.38 / .53
Thin-net nonplenum IEEE 802.3 10-Base-2	6417	1354 80°C	19/.0072 Tin Copper	Solid PE .116	T.C. Braid—95% Coverage	PVC Black .195	66%	50	30.8	5	10	.99 / 1.4
Thin-net nonplenum IEEE 802.3 10-Base-2	6418	1354 80°C	19/.0072 Tin Copper	Foam PE .094	Alum/Tape + T.C. Braid 95% Coverage	PVC Black .185	78%	50	26.0	5	10	.99 / 1.4
Thin-net plenum IEEE 802.3 10-Base-2	6420	NEC 725-38 (b) (3) 125°C	19/.0072 Tin Copper	Solid FEP[1] .110	Alum/Tape + T.C. Braid 95% Coverage	Kynar Flex[3] Black Tint .174	66%	50	30.8	5	10	.99 / 1.4
Thin-net plenum IEEE 802.3 10-Base-2	6419	NEC 725-38 (b) (3) 125°C	19/.0072 Tin Copper	Foam FEP[1] .090	Alum/Tape + T.C. Braid 95% Coverage	Kynar Flex[3] Black Tint .160	78%	50	26.0	5	10	.99 / 1.4

[1]FEP = Registered trademark of de Nemours & Co.

[2]Trademark of Xerox Corp.

[3]Kynar—Registered trademark of Penwalt Corporation

Note: All constructions shown can be altered to satisfy individual customer requirements. For instance, plenum jacketing materials can be FEP, Halar, etc. Consult Graybar with specific requirements.

Intelligence Building Composite Cables

	Description	MONCO CBL	UL Style	Conductor Material Stranding Diam (Inch)	Dielectric Material Diameter (Inch)	Shield Material % Coverage	Jacket Material Nom Diam (Inch)	Nom Imped (ohms)	Nominal Attenuation MHz dB/100'
Figure 1	34/c Mini RG-62 plenum	5779-34	NA	26 AWG Solid Silver Plated Copper	Foam FEP .104	T.C. Braid 90% coverage	Fluoropolymer clear 1.011	95	NA
Figure 2	RG-6 type + 4-pair telephone composite plenum	0585-04	NA	A. 18" AWG Bare / B. 24 AWG Bare	A. Foam FEP .180 / B. Halar .031	A. Alum. Mylar + T.C. Braid / NA	A. FEP / NA / Overall Jacket FEP	75 / 100	50 1.5 / 100 2.1 / NA NA
Figure 3	36/c composite	6459	NEC 725-38 (b) (3)	A. #24 AWG Solid Bare / B. #22 AWG Solid Bare / C. #22 Solid Bare	A. Halar[2] .034 / B. Halar[2] .040 / C. FEP[1] .068	A. No Shield / B. No Shield / C. Alum/Tape with #24 T.C. Drain Wire	Fluoropolymer Black .532		
Figure 4	Voice, data, and video composite	6433	NEC 725-38 (b) (3)	C. #24 AWG Solid Bare Copper / D. #22 Solid Bare Copper / E. #22 Solid Bare Copper Covered Steel	C. Solid FEP[1] .037 / D. Solid FEP[1] .037 / E. Foam FEP[1] .103	C. Overall Tape and #24 AWG Solid Drain / D. Individual Alum/Tape and #24 AWG solid drain / E. T.C. Braid 95% Coverage	C. Fluoropolymer Clear .204 each pair / D. Fluoropolymer clear .226 each pair / E. Fluoropolymer .157		Overall Jacket Fluoropolymer .486"

1 = Trademark of de Nemours & Co.
2 = Trademark of Allied Corporation
Note: Please consult Graybar for designs not shown here.

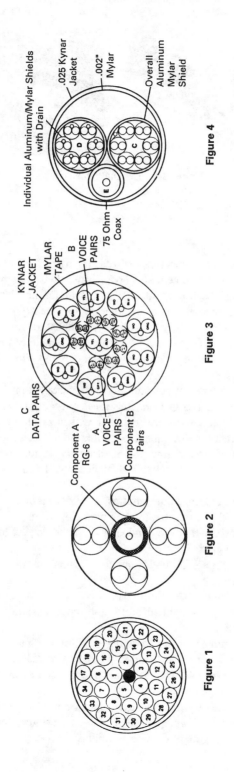

.025 Kynar Jacket

.002" Mylar

Overall Aluminum Mylar Shield

Individual Aluminum/Mylar Shields with Drain

D

C

E

75 Ohm Coax

Figure 4

KYNAR JACKET

MYLAR TAPE

B VOICE PAIRS

C DATA PAIRS

A VOICE PAIRS

Component A RG-6

Component B Pairs

Figure 3

Figure 2

Figure 1

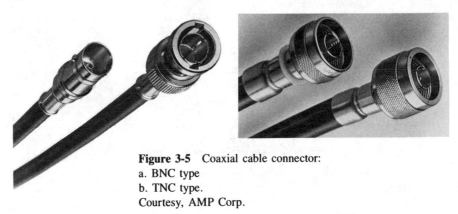

Figure 3-5   Coaxial cable connector:
a. BNC type
b. TNC type.
Courtesy, AMP Corp.

The two types of connectors utilized are named Bayonet Neill-Concelman (BNC) and Twist Neill-Concelman (TNC) after their designers. The BNC is a bayonet type that has been used since World War II. The TNC is a screw-on type that has been developed for easy installation to video-type cable (those with a solid shield). Examples of these two types of connectors are shown in Figure 3-5.

Coaxial cable can be terminated with either a male or female connector, and various adapters can be connected to a coaxial cable to allow for parallel or series connections (Figure 3-6).

However, three types of connectors can be installed on to coaxial cable (Figure 3-7). These are the **crimp type,** the **three piece,** and **screw-on type;** crimp-type connectors (Figure 3-7a) are installed with a crimping tool and a wrench. The advantage of this type of connector is that no soldering is required, eliminating the possibility of overheating and melting the inside insulation between the center conductor and the ground shield, which could cause a short circuit as the connector is fitted or in the future.

The three-piece connector requires a soldering iron to solder the center probe that extends the conductor to the connector. Care must be taken when assembling this connector that the conductor is of the correct length, the pin is not overheated, and the ground braid (shield) is correctly installed in the holding shell and securing nut. Figure 3-7b depicts the correct installation methods for the three-piece connector.

Figure 3-7c depicts the crimp coaxial connector. To attach this type of connector, the outer insulation, shield braid, and inner insulation are trimmed with a coaxial cable stripper.

Table 3-4 summarizes the advantages of the two-piece connector requiring only one crimping action. The detailed procedure for cable preparation with this tool is shown in Figure 3-8. This tool can also be used for preparing the cable for the other types of connectors.

TNC type cable requires that a crimper be used to secure the body of the connector to the shield. Figure 3-9 depicts such a crimper that also may be used for connecting an assortment of wire terminals.

Figure 3-6 Examples of coaxial cable connector adapters.

Terminating instructions for BNC solder connectors

Nut   Washer   Gasket   Clamp   Bushing   Male Contact   Plug body

Don't use bushing with RG 58 A/U

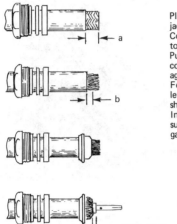

→| |←— a

→| |←— b

Place nut, washer and gasket over cable and cut jacket to dimension shown.
Comb out braid and fold out. Cut cable dielectric to dimension. Tin center conductor.
Pull braid foward and taper toward center conductor. Place clamp over braid and push back against cable jacket.
Fold back braid wires as shown, trim to proper length (1/8", 3.2 mm.) and form over clamp as shown. Solder contact to center conductor.
Insert cable and parts into connector body. Make sure sharp edge of clamp seats properly in gasket. Tighten nut.

→| |←— 3/32 (2.4)

(a)

Terminating instructions for BNC crimp connector

Plug body assembly

Outer ferrule   Mil-crimp

| a | b | c |

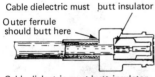

Contact must butt against cable dielectric

Cable dielectric must butt insulator

Outer ferrule should butt here ——→

Cable dielectric must butt insulator

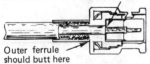

Outer ferrule should butt here

Strip cable jacket, braid, and dielectric to dimensions shown. Stripping dimensions: a. .34375": b. .09375": c. .15625". All cuts are to be sharp and square.
Important: Do not nick braid, dielectric, and center conductor. Tinning of center conductor is not necessary if contact is to be crimped. For solder method, tin center conductor avoiding excessive heat. Slide outer ferrule onto cable as shown. Flare slightly end of cable braid as shown to facilitate insertion ontoi inner ferrule.
Important: Do not comb out braid. Place contact on cable center conductor so that it butts against cable dielectric. Center conductor should be visible through inspection hole in contact. Crimp or solder the contact in place as follows:

Crimp method: Use Cavity B of die set
Solder method: Soft solder contact to cable center conductor. Do not get any solder on outside surfaces of contact. Avoid excessive heat to prevent swelling of dielectric.

Install cable assembly into body assembly so that inner ferrule portion slides under braid. Push cable assembly forward until contact snaps into place in insulator. Slide outer ferrule over braid and up against connector body. Crimp outer ferrule using Cavity A of die set.

(b)

Figure 3-7   Examples of coaxial cable connectors:

a. crimp type

b. three-piece type

c. screw-on connector.

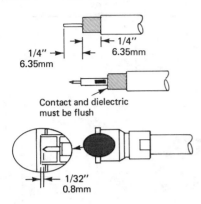

1. Using the Paladin ETN wire stripper or suitable alternate, prepare the end of the coaxial cable for connector instsllation.
2. Trim cable as shown, taking care not to nick the center conductor or outer braid.
3. Ensure that the outer braid lays flat.
4. Trim off any and all excess braid.
5. Twist the contact in a clockwise direction, on the inner conductor, untill the back end of the contact is flush with the inner dielectric.
6. Twist the connector onto the cable in a clockwise direction. The connector is properly installed when the end of the contact is positioned within 1/32" of the front edge of the connector.

(c)

When installing a workstation the connecting coaxial cable is run **home** to the source equipment (Figure 3-10a) or is **daisy chained** with other devices within the computer/data network (Figure 3-10b). Regardless of the type of connection to the network, the cabling from the workstation will probably connect to a wall plate or to a patch panel (Figure 3-11) in a wiring closet or wiring a cabinet. The workstations may be terminated to output to twisted-pair cabling, in which case the output impedance of the device would differ from that of the coax. The two impedances must be matched by an impedance matching device called a **balun.** The balun can be connected between the workstation and the face plate between the two different wiring media or at a patch panel. There are a multitude of choices for balun connections; Figure 3-12 depicts several types.

Figure 3-13 illustrates how coaxial adaptors might be used to connect cables. The coaxial adaptors used in this illustration are BNC type. A splice insulating cover is used to protect the connection to the ''T'' type connector.

TABLE 3-4   The Advantages of the Two-Piece Connector
Over the Three-Piece Connector.

- Two parts make up a complete plug or jack
- No danger of heat damage to coaxial cable
- Fully intermateable with comparable UG/U Series Connectors
- Improved cable retention and insulation grip
- Ease of inspection
- Stabilized inner contacts
- Less critical stripping dimensions than required for solder assemblies
- Low VSWR
- Reduced noise level
- Simplified replacement in field
- Full responsibility for the development and performance of the crimp
- Positive insulation grip with crimped braid ferrule
- Lightweight—¾ ounce (23.33 gm) (cable plug and cable jack)

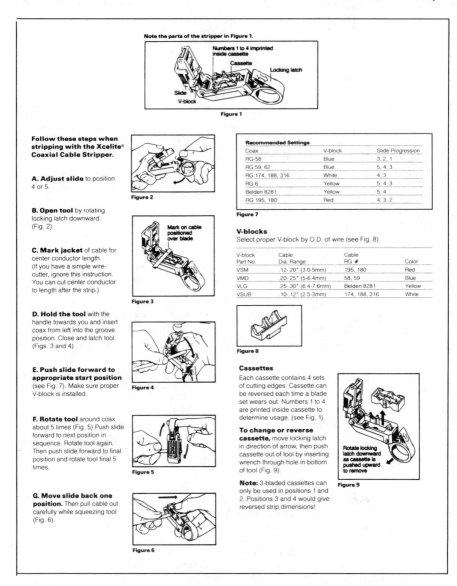

Note the parts of the stripper in Figure 1.

Numbers 1 to 4 imprinted
inside cassette

Cassette

Locking latch

Slide

V-block

Figure 1

**Follow these steps when
stripping with the Xcelite®
Coaxial Cable Stripper.**

A. Adjust slide to position
4 or 5.

Figure 2

B. Open tool by rotating
locking latch downward.
(Fig. 2)

C. Mark jacket of cable for
center conductor length.
(If you have a simple wire-
cutter, ignore this instruction.
You can cut center conductor
to length after the strip.)

Mark on cable
positioned
over blade

Figure 3

D. Hold the tool with the
handle towards you and insert
coax from left into the groove
position. Close and latch tool.
(Figs. 3 and 4)

**E. Push slide forward to
appropriate start position**
(see Fig. 7). Make sure proper
V-block is installed.

Figure 4

F. Rotate tool around coax
about 5 times (Fig. 5) Push slide
forward to next position in
sequence. Rotate tool again.
Then push slide forward to final
position and rotate tool final 5
times.

Figure 5

**G. Move slide back one
position.** Then pull cable out
carefully while squeezing tool
(Fig. 6).

Figure 6

Figure 7

Recommended Settings

Coax	V-block	Slide Progression
RG-58	Blue	3, 2, 1
RG 59, 62	Blue	5, 4, 3
RG 174, 188, 316	White	4, 3
RG 6	Yellow	5, 4, 3
Belden 8281	Yellow	5, 4
RG 195, 180	Red	4, 3, 2

V-blocks
Select proper V-block by O.D. of wire (see Fig. 8).

V-block Part No.	Cable Dia. Range	Cable RG #	Color
VSM	.12-.20″ (3.0-5mm)	195, 180	Red
VMD	.20-.25″ (5-6.4mm)	58, 59	Blue
VLG	.25-.30″ (6.4-7.6mm)	Belden 8281	Yellow
VSUB	.10-.12″ (2.5-3mm)	174, 188, 316	White

Figure 8

Cassettes
Each cassette contains 4 sets
of cutting edges. Cassette can
be reversed each time a blade
set wears out. Numbers 1 to 4
are printed inside cassette to
determine usage. (see Fig. 1)

**To change or reverse
cassette,** move locking latch
in direction of arrow, then push
cassette out of tool by inserting
wrench through hole in bottom
of tool (Fig. 9).

Note: 3-bladed cassettes can
only be used in positions 1 and
2. Positions 3 and 4 would give
reversed strip dimensions!

Rotate locking
latch downward
as cassette is
pushed upward
to remove

Figure 9

Figure 3-8   Coaxial wire strippers with details for operation. Courtesy,
Xcelite, Inc.

3.4 Grounding of Coaxial Cables

In Chapter 2 we discussed the necessity of grounding, the problems that can occur
without proper grounding, and the results of ground loops in the grounding of a
cable. A review of Section 2.3 (Shielding of Transmission Lines) may be in order.

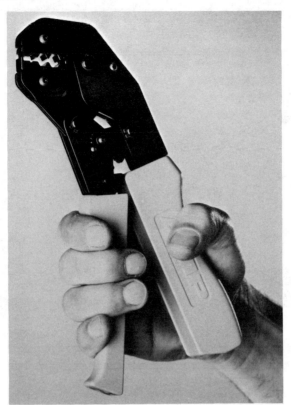

Figure 3-9   An example of a mechanical crimping tool. Courtesy, Amp Corp.

In this chapter we discuss only the grounding of coax as it pertains to the National Electrical Code. Figure 3-14 depicts the proper grounding techniques for coax that is run through metal conduit, and into a service enclosure. The cable shield is grounded by a ground clamp to the bonding bushing of the conduit. The earth grounding conductor must be less than 15 meters (50 feet) in length and AWG 2-8 solid copper; insulation color is to be green or green and yellow. Barrel connectors should be installed at the entrance of each building entry point and all exposed connectors should be insulated as protection from the weather.

The neutral bar on the inside of the service box is connected to the grounding bushing of the conduit, and is in turn connected to a ground electrode. The grounding electrode conductor is typically of #2 to #8 AWG bare copper wire and must be run from the box through a cable bushing and conduit to the grounding electrode.

The conduit for the grounding conductor can usually be of EMT-type plastic. The grounding rod is a bronze rod driven at least 10 feet into the earth or grounding conductors embedded in the building foundation.

Local codes should be checked to assure compliance.

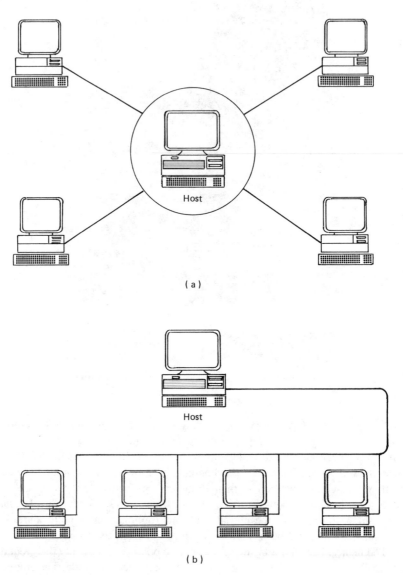

(a)

(b)

Figure 3-10 Examples of connections of workstations to a Host computer:
a. home run
b. daisy chained.

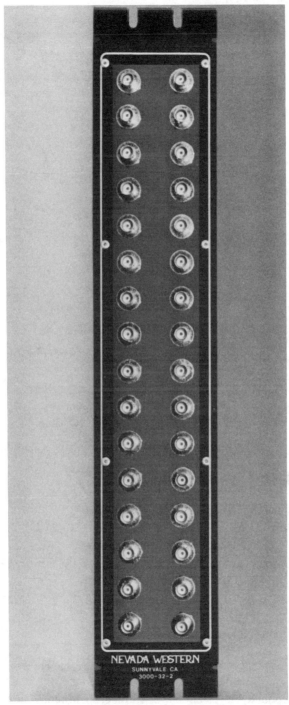

Figure 3-11 An example of a patch panel. Courtesy, Nevada Western, Inc.

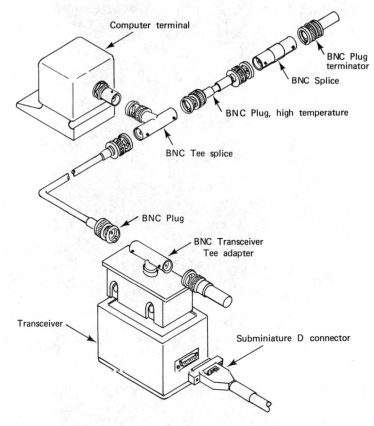

Figure 3-12 Applications of connectors. Courtesy, Amp Corp.

3.5 Applications of Coaxial Cables

1. Closed circuit television
2. Cable television
3. Video security systems
4. Computer systems
5. Communication systems

3.6 Advantages of Coaxial Cables

The characteristics that are an advantage for coaxial cable are

1. low susceptibility to electromagnetic pick up resulting in less noise and cross-talk

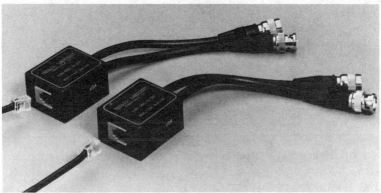

Figure 3-13 Examples of balun matching devices:
a. RJ-type connector to coax
b. patch panel connector
c. breakout box.
Courtesy, Amp Corp.

2. high bandwidth for transmitted signals, resulting in low signal distortion
3. utilization over longer distances than twisted-pair cables
4. proven performance and reliability over many years
5. can be matched to operate with both fiber-optic and twisted-pair systems
6. a lower signal distortion due to phase shift, or variation of amplitude with increased frequency
7. much larger number of channels can be transmitted over the same cable
8. less cross-talk between cables
9. greater information security than twisted pairs

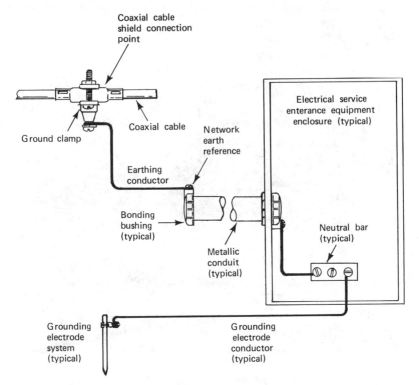

Figure 3-14   Proper grounding of a coaxial entrance cable.

3.7 Disadvantages of Coaxial Cables

The characteristics of coaxial cable that are a disadvantage as compared to twisted pair are

1. more difficult to install than twisted pair
2. heavier than twisted-pair or fiber-optic cables
3. many systems have shifted from coaxial cable to twisted pair as new technology is developed to improve the transmission along twisted pair
4. usually must be daisy chained or home run to workstations
5. doesn't have the adaptability of twisted pair
6. is more expensive and takes more time to install than twisted pair

3.8 Applications of Coaxial Cable

1. Closed circuit cable
2. Cable TV

3. Video security systems
4. Computer systems
5. Communication systems

Summary

Coax is a time-proven cabling that can be utilized in most audio/data communication systems and with most telecommunication devices. The additional expense of coaxial cable is usually justified over twisted-pair cable if information security and signal bandwidth are important. Coaxial cable is less expensive to install than fiber-optic cable.

Questions

1. What are the advantages of coaxial cable over twisted pair?
2. When would the additional cost of coax over twisted pair be justified?
3. What is the purpose of a balun being connected between a workstation and a coaxial cable?
4. What is the purpose of placing a balun between a coaxial cable and a twisted-pair cable?
5. What is the major disadvantage of coax over twisted-pair wiring?

CHAPTER 4

Fiber Optics

4.1 Introduction

The science of fiber optics deals with the transmission of electromagnetic energy in the form of light along a transparent medium such as glass or plastic. When one of these materials is formed into a transparent wire, and coated by a material that is less light conductive, an effective mirror is formed around the fiber. This mirror effect produces a **light pipe** through which light can be transported.

Light rays introduced into the fiber travel down the clear medium, of the light pipe, reflecting back and forth from the edges of the fiber (Figure 4-1). The angle of reflection (leaving the mirror surface) is the same as the angle of refraction (entering the mirror surface). If the refraction angle is too great, the light will not be reflected and will travel through the cladding much as light rays from the sun reflect or enter the surface of a lake at different times of the day. The angle at which light rays will not be reflected from a surface is called the **critical angle.** We need not concern ourselves further with this phenomenon.

Fiber-optic cable is constructed with a small center transparent core which is covered by the cladding and sealed by an outer jacket. The cladding, a less transparent glass than the fiber, protects the fiber and aids in the reflection process. The jacket is a tough plastic covering that protects both the cladding and the fiber from the outside environment: heat, moisture, dirt, scratches, and nicks.

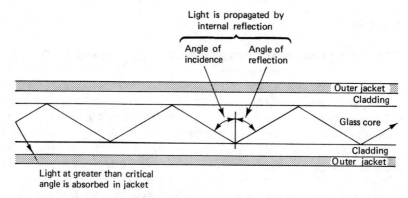

Figure 4-1 Light waves reflecting back and forth along the inside of a fiber.

Figure 4-2 depicts a cutaway view of a fiber optic cable showing the transparent core, the cladding, and the jacket.

Fiber optics, the technology for the transmission of the visible light portion of the electromagnetic frequency spectrum through a glass fiber, has been a technology waiting for a widespread application. The need for wiring systems with greater signal capacity, greater bandwidth, and information security in the telecommunication industry is the application that has expanded the use of fiber-optic cables.

Fiber cables, as compared to copper cables, have greater signal capacity, are unaffected by electromagnetic waves, create no electromagnetic radiation, and therefore, offer *extremely good information security.* It is almost impossible to "steal information" from a fiber cable without being detected.

As we stated earlier, fiber-optic cables transmit signal information with light waves as the signal carrier and special glass or plastic fibers as the transmission medium. The transmission of light through a fiber is a rather complex science; a complete treatment is beyond the scope of this text, and it is not needed for the installation and application of such cable.

We will discuss the characteristics of fiber light transmission that should be helpful in selecting a fiber type and understanding the installation procedures.

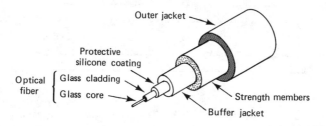

Figure 4-2 A cutaway view of a fiber-optic cable showing the transparent core, the cladding, and the jacket.

4.2 Fiber Types

Optical fibers are classified by material composition, the refraction index of the core, and the modes (signal transmission methods) of the fiber. The **refraction index** is the angle that the light is reflected from the walls of the fiber. The **mode** is the process of how the light waves are propagated (transmitted) through the fiber. The composition of the fiber, although usually glass, can be plastic.

Light in the form of photons of energy is transmitted (propagated) down the fiber by what are called modes. Mode is a concept describing the way the light waves travel through a medium. James Clark Maxwell, a Scottish physicist of the last century, showed that electromagnetic waves were a single form of energy and that propagation followed strict rules, now known as Maxwell's law. For our purpose, mode is simply the path that light travels down a fiber (Figure 4-3). The number of modes that can be transmitted down a fiber cable range from one for a single fiber cable to over a million for a multimode cable configuration.

The diameter of the fiber determines if a fiber is a single mode or multimode. When the diameter of a fiber is reduced to 5 from 10 micrometers (μm), it will support only one mode of light transmission. The standard cladding for a single-mode cable is 125 μm. This value was chosen because:

1. It is the same size as graded-index fibers, allowing a size standardization.
2. The cladding must be approximately 10 times or greater than the core.
3. It will make the fiber cable less brittle and easier to handle during installation.

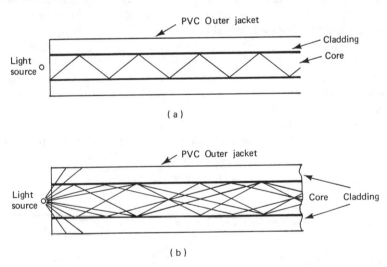

Figure 4-3   An illustration of a light signal transmission along a single-mode and a multimode cable:
a. single mode
b. multimode.

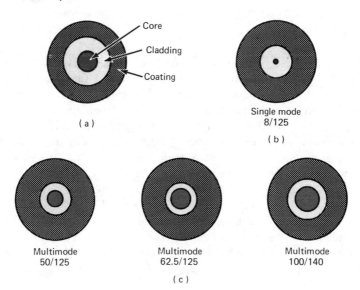

Figure 4-4   An illustration of different types of fiber cables:
a. crosssection of fiberoptic cable
b. single mode
c. three sizes of multimode cable

Single-mode fibers (Figure 4-4b) have a bandwidth potential of over 50 MHz. Generally, the bandwidth of a system is limited by the electronic devices within the system and not by any of the fiber-optic cables. The advantage of single-mode cabling is that it allows for future additions and improvements. The characteristics of light propagating through a fiber cable depends upon factors such as the composition of the fiber, the size of the fiber, and the light optically injected into the fiber.

Multimode fiber cable (Figure 4-4c) is classified as multimode-step-index fiber (step-index fiber) and multimode-graded-index fiber (graded-index fiber).

Step-index fiber is a multimode fiber of glass or plastic and is constructed with a core diameter from 100 to 970 μm. Although step-index fiber is not the most efficient it is the most often used. Step-index fiber experiences modal dispersion or the spreading of light with transmission within the fiber. This is because light reflects at different angles for different modes (paths) and, therefore, takes different times to reach a given point in the fiber.

Most graded-index fibers have a core diameter of 50, 62.5, or 85 μm and the cladding diameter is 125 mm. This fiber is used in applications such as telecommunications that require a large bandwidth. AT&T has adopted for its standard a diameter of 62.5 mm, and the IEEE has proposed this standard. Because of these actions 62.5 mm will probably become an industry standard.

4.3 Cable Construction

Fiber-optic cables are constructed in special forms to best serve the application to which they are placed. However, all are comprised of a fiber light conductor, a light reflecting cladding, and an outer jacket. The cladding protects the fiber and acts as a reflector of the light to prevent the light below the critical frequency from escaping from the fiber and also seals the light in the cable.

The function of the outer coating is to protect the cladding and the fiber. The outer coating, comprised of one or more layers of polymer (plastic), protects the interior from moisture and physical abuse, such as dirt and nicks.

Figure 4-5 is a summary of several packing techniques for fiber cable. Figure 4-5a shows a single fiber called simplex fiber. While a single fiber (Figure 4-5a) will carry thousands of separate data streams, it has the disadvantage in that a break would put the entire system out of service. Figure 4-5b depicts a cable with two separate fibers. This cable has twice the capacity of the simplex cable and the advantage of a backup cable. The zip cable in Figure 4-5d is a dual cable that can

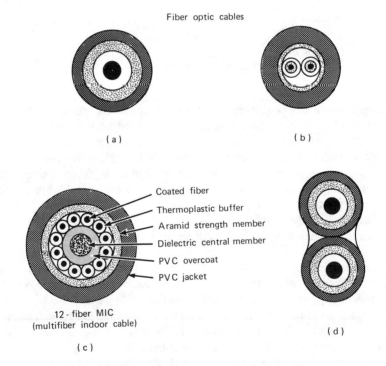

Fiber optic cables

(a)

(b)

Coated fiber
Thermoplastic buffer
Aramid strength member
Dielectric central member
PVC overcoat
PVC jacket

12 - fiber MIC
(multifiber indoor cable)

(c)

(d)

Figure 4-5 Packaging methods for fiber cable:
a. simplex fiber
b. dual fiber
c. multiple fiber with a strength member
d. dual zip cord cable.

easily be separated. It is flatter than the other type of dual cable. Figure 4-5c shows a multicable with a strong center strength member that protects the fiber from stress and strain. This cable probably has more conductors than necessary for most applications. However, the designer should plan on an order of three times the base requirements for expansion.

A strength member can be included in any type of cable and should always be included in a cable that is to be pulled through a conduit or hung between buildings.

Technically speaking, the entire makeup of conductor and covering is referred to as a cable. However, here we will refer to all the outer covering that protects one or more internal fibers from the stress and strain of the external environment as cabling.

A buffered jacket, which protects the fiber and the cladding, is put on the cable by the fiber manufacturer. Additional buffering material, the strength member, and the outer jacket is usually put on the cable by the cable manufacturer. These materials are chosen to protect the cable in a particular environment. Two techniques are used for this final buffering: loose buffering and tight buffering. When tight buffering is utilized, the buffering is applied tightly over the cladding and the fiber. This forms a solid tough cable which has the advantage of flexibility for tight radii in bends with low losses. Tight buffering has the additional advantage of high impact and crush resistance. The disadvantages will be discussed later in this chapter.

Loose buffering (Figure 4-6) is formed by adding a loose covering, much like a sleeve, over the cladding. More than one fiber may be included in the buffering. The advantage of loose buffering is that the buffer tube is designed to take any stress or strain placed on the cable and relieve the fiber from change of length due to retraction or expansion with a change of temperature.

Strength members may be made of steel, fiber glass, or Aramid^{TM} yarn. Steel, while the strongest, should not be used in all optical cables. However, it offers a very strong support for cables that must be strung between buildings or poles.

The small size of the fiber and the cladding allows cables to be formed into almost any special configuration. For example, the cable in Figure 4-7a which is only 0.07 inch thick would hardly make a ripple under a carpet. Another unique application is the very thin ribbon cable in Figure 4-7b that has an outer coating which allows greater flexibility.

4.4 Cable Characteristics

As was stated earlier, dispersion is the spreading of a light pulse as it travels along the optical cable. Dispersion limits bandwidth and information-carrying capabilities of a cable. The bit rate can be high enough so that the pulses overlap and become unintelligible. Figure 4-8 illustrates the problem of pulse spreading.

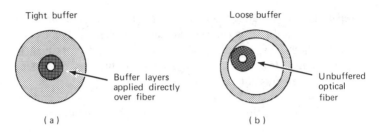

Tight buffer   Loose buffer

Buffer layers
applied directly
over fiber

Unbuffered
optical
fiber

(a)   (b)

Cable parameter	Cable structure	
	Loose tube	Tight buffer
Bend radius	Larger	Smaller
Diameter	Larger	Smaller
Tensile strength installation	Higher	Lower
Impact resistance	Lower	Higher
Crush resistance	Lower	Higher
Attenuation change at low temperatures	Lower	Higher

(c)

Figure 4-6 Comparison of types of buffering:
a. loose buffered cable
b. tight buffered cable
c. comparison of the characteristics of the two types.
Courtesy, Belden Corp.

Dispersion or spreading for step indexed fibers is 15 to 30 ns/km. This means that two modes entering a fiber can be separated by 30 nanoseconds in a 1000 meter cable. Eventually, two pulses could combine to form an unintelligible pulse.

Graded-index fiber has a core of numerous concentric layers of glass. Each layer of the core refracts the light. Instead of being sharply refracted, as in the step-index fiber, the light is bent in an almost sinusoidal pattern. The light near the center has the lowest velocity, and the light near the outer rings has the greatest velocity. This produces a more even travel rate and reduces the pulse spreading effect.

The bit rate of the pulses must be low enough so as not to overlap due to dispersion. There are three main types of dispersion in fibers: modal, material, and waveguide. However, we will not concern outselves with the specific details of each of these since cable manufacturers do not differentiate between the types. Cable manufacturers do not specify a dispersion factor for multimode cables. Instead a **figure of merit** called the **bandwidth length product** or simply **bandwidth** is given in megahertz per kilometer. A 500 MHz/km designation means that a 500 MHz signal can be reproduced after a transmission of 1 km or a 250 MHz signal can be reproduced after a transmission of 2 km. For example:

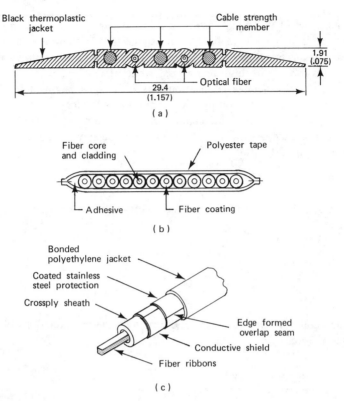

Figure 4-7   Examples of special cable configurations for fiber-optic cables:
a. under the carpet
b. ribbon cable
c. cable construction.

$$1000 \text{ MHz} - 0.5 \text{ km}$$
$$500 \text{ MHz} - \quad 1 \text{ km}$$
$$250 \text{ MHz} - \quad 2 \text{ km}$$
$$100 \text{ MHz} - \quad 5 \text{ km}$$

The dispersion factor for single-mode cable is expressed in picoseconds per kilometer per nanosecond (ps/km/nsec) of source spectral width. This means that in single-mode cable, the depression factor is most affected by the source spectrum (frequency or light). The wider the source spectrum or the greater the number of wavelengths of the source light, the greater the attenuation.

Input pulses   Glass fiber   Output pulses

Figure 4-8   Illustration of pulse spreading in fiber-optic cable.

Attenuation of the signal results in a loss of signal power. Attenuation in a fiber-optic cable is the loss of light as the signal travels along the cable. Attenuation varies as the wavelength or frequency of the light varies. Some wavelengths or frequencies travel with little or no loss, while others have a large loss. Figure 4-9 shows a chart of decibel attenuation versus the wavelength of light over the wavelength of 700 mm to 1600 mm for a multimode cable.

Figure 4-10 shows the attentuation characteristics of a single-mode cable over the approximate wavelength. Note that a frequency of 200,000 megahertz has a wavelength of 1.5μm.

$$\lambda = \frac{C}{f}$$

where C = speed of light (186,000 miles/sec) and f = frequency in cycles per second (Hz)

$$\lambda = \frac{300}{f} \text{ meters}$$

$$\frac{300}{2} \times 10 = 1500 \text{ mm}$$

When long cable runs are necessary one can study the charts and avoid the high-loss frequency areas.

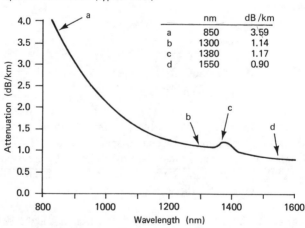

Characterization data

Characterized parameters are typical values.

Core material index of refraction (peak):   1.4805 at 850 nm
                                              1.4748 at 1300 nm
Spectral attenuation (typical fiber):

	nm	dB/km
a	850	3.59
b	1300	1.14
c	1380	1.17
d	1550	0.90

Figure 4-9 Attenuation in dB per km versus wavelength over the 0.8 to 1.6 μm range. Courtesy, Corning Glass Works.

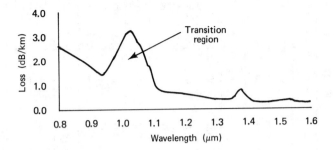

Figure 4-10   A chart of the attenuation in dB per km versus wavelength for a single-mode cable. Courtesy, Corning Glass Works.

Of all the advantages of fiber-optic cables over copper cables, the most important is perhaps that within the bandwidth of a signal mode the loss in the fiber cable is constant.

Table 4-1 depicts the operating characteristics of fiber optic cable. The table gives the attenuation, maximum length, bandwidth, size and weight, operation temperature, tensile load during installation, crush resistance, and flame rating for each size of cable.

4.5 Cable Terminations

Energy transmission through a fiber-optic cable depends upon the passage of light through the fibers. Any restriction to this passage blocks some of the light and increases signal attenuation. For this reason special care must be taken when attaching connectors and splicing the cable. Splicing is usually required in very long cables because cable is easier to purchase in the 1 to 5 km length. It is also very difficult to pull a cable of, say, 20 km.

Splicing and connectors are necessary when adding new cable to an existing system, connecting exterior to interior cable, splitting cable information into several work positions, connecting the cable to a transmitter or receiver, breaking the system into several subsystems, and connecting the system into a switching system.

Forming the fiber into several subsystems allows for ease of service, changing equipment as personnel change, and upgrading equipment without disturbing the system.

There are many connector manufacturers that produce a number of different types of connectors. Figure 4-11 depicts three industry standards. Adapters can be purchased to interconnect any of these connector types. The biconic type is recommended for single mode, and the ST is recommended for multimode operation. All these connectors can be used with adapters which can be purchased to adapt one type connector to the other.

TABLE 4-· Nevada Western

Fiber Size (μm)	Description	Atten. Max. 850 nm (dB/km)	Atten. Max. 1300 nm (dB/km)	Bandwidth Min. 850 nm (MHz-km)	Bandwidth Min. 1300 nm (MHz-km)	Dim. Nom. (mm)	Cable Weight (kg/km)	Operating Temp. (°C)	Tensile Load Install. (N)	Bond Radius Min. @ Install. (cm)	Crush Resistance (N/cm)	Flame Rating[1]	Part Number
50/125	Light Duty Single, PVC	4.0	2.5	400	400	3.0	9.0	−20/+80	420	4.0	550	OFNR	502082-1[2]
	Heavy Duty Single	4.0	2.5	400	400	3.7	13.0	−20/+70	560	8.0	550	OFNR	501740-1[2]
	Light Duty Dual, PVC	4.0	2.5	400	400	3.0x6.1	18.0	−20/+80	840	4.0	550	OFNR	502085-1[2]
	Heavy Duty Dual	4.0	2.5	400	400	3.8x7.8	26.5	−20/+70	1120	8.0	550	OFNR	501116-5[2]
	Plenum Grade Single	4.0	2.5	400	400	2.8	6.5	−20/+80	420	5.0	450	OFNP	501819-1
	Plenum Grade Dual	4.0	2.5	400	400	2.8x6	13.1	−20/+80	840	5.0	450	OFNP	501693-1
	Breakout-2 Fiber	3.5	2.5	500	500	8.5	55.0	−30/+70	1800	17.0	700	—	501438-2
	HD Breakout-2 Fiber	3.5	2.5	500	500	9.4	70.0	−30/+70	4300	19.0	700	—	501498-2
62.5/125	Light Duty Single, PVC	4.0	2.5	160	500	3.0	9.0	−20/+80	420	4.0	550	OFNR	502083-1[2]
	Heavy Duty Single	4.0	2.5	160	500	3.7	13.0	−20/+70	560	8.0	550	OFNR	501739-1[2]
	Light Duty Dual, PVC	4.0	2.5	160	500	3.0x6.1	18.0	−20/+80	840	4.0	550	OFNR	502086-1[2]
	Heavy Duty Dual	4.0	2.5	160	500	3.7x7.8	26.5	−20/+70	1120	5.0	550	OFNR	501738-1[2]
	Plenum Grade Single	4.0	2.5	160	500	2.8	6.5	−20/+80	420	5.0	450	OFNP	501820-1
	Plenum Grade Dual	4.0	2.5	160	500	2.8x6	13.1	−20/+80	840	5.0	450	OFNP	501754-1
	DUALAN	4.0	1.5	160	500	4.75	20.0	−20/+80	1000	10.0	700	OFNR	501749-1[2]
	DÚALAN Plenum	4.0	1.5	160	500	4.75	20.0	−20/+80	1250	10.0	700	OFNP	502024-1

Fiber Size	Cable Type												Flame Rating	Part Number
	Breakout-2 Fiber	4.0	2.5	160	500	8.5	55.0	−30/+70	1800	17.0	700	—	501438-4
	HD Breakout-2 Fiber	4.0	2.5	160	500	9.4	70.0	−30/+70	4300	19.0	700	—	501498-4
	IBM Type 5	6.0	4.0	150	500	7.5	48.0	−20/+80	1000	15.0	*	—	501714-1
	Light Duty Single, PVC	5.0	4.0	100	200	3.0	9.0	−20/+80	420	4.0	550	OFNR	502084-1[2]
	Heavy Duty Single	5.0	4.0	100	200	3.7	13.0	−20/+70	560	8.0	550	OFNR	501741-1[2]
100/140	Light Duty Dual, PVC	5.0	4.0	100	200	3.0x6.1	18.0	−20/+80	840	4.5	550	OFNR	502087-1[2]
	Heavy Duty Dual	5.0	4.0	100	200	3.8x7.8	26.5	−20/+70	1120	8.0	550	OFNR	501118-5[2]
	Plenum Grade Single	5.0	4.0	100	200	2.8	6.5	−20/+80	420	5.0	450	OFNP	501821-1
	Plenum Grade Dual	5.0	4.0	100	200	2.8x6	13.1	−20/+80	840	5.0	450	OFNP	501755-1
	Breakout-2 Fiber	4.0	3.0	200	300	8.5	55.0	−30/+70	1800	17.0	700	—	501438-6
	HD Breakout-2 Fiber	4.0	3.0	200	300	9.4	70.0	−30/+70	4300	19.0	700	—	501498-6
Single-Mode	Light Duty Single, PVC	—	1.0	—	—	3.0	9.0	−20/+80	250	4.0	550	OFNR	501530-1[2]
	Breakout-2 Fiber	—	1.0	—	—	8.5	55.0	−30/+70	1050	17.0	700	—	501556-1
	DUALAN	—	1.0	—	—	4.8	20.0	−20/+80	1250	10.0	700	OFNR	502119-1[2]
	DUALAN Plenum	—	1.0	—	—	4.8	20.0	−20/+80	1250	10.0	700	OFNP	502120-1

Flame Ratings—Flame ratings for cables used within buildings are specified by the National Electrical Code (NEC). Underwriters Laboratories (UL), through their listing process, defines procedures to assure that products meet specific requirements.

General Use (OFN)—These cables pass the UL 1581 Vertical Tray Flame Test. They may be used for general installation in building wiring. They may not be used in risers or plenums unless installed in suitable conduits.

Riser Cable (OFNR)[2]—Riser Cables must meet requirements of UL 1666. and may be used in vertical passages connecting one floor to another.

Plenum Cable (OFNP)—Plenum cable must pass the Steiner Tunnel Test, UL 910. The cable may be installed in air plenums without the use of conduit.

Note: In all cases, cables meeting the more stringent requirements may be used in place of a particular cable.

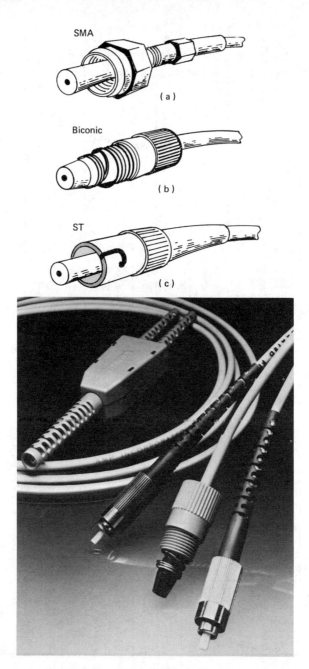

SMA

(a)

Biconic

(b)

ST

(c)

Figure 4-11 Types of fiber-optic connectors:
a. SMA
b. Biconic
c. St
d. Photos.
Courtesy Amp Corp.

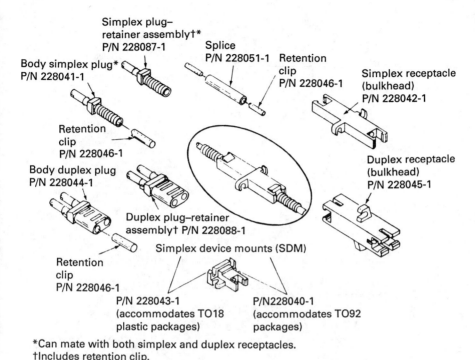

*Can mate with both simplex and duplex receptacles.
†Includes retention clip.

Figure 4-12   Types of P/N fiber connectors. Courtesy, Amp Corp.

The biconic type is most often used for single-mode operation and the ST type is most often used for multimode application. Figure 4-12 depicts several P/N types of connectors.

The features of an ideal connector are that it is economical, has low loss, is easily installed, and has reliability. A connector must be economical to install. This means that installation can be accomplished quickly and reliably by employees with a minimum of training. A connector should present a minimum attenuation (dB) loss to the signal. A connector should perform reliably and consistently after many connects and disconnects. The initial cost is probably the least important consideration for purchasing a given type of connector.

Losses in an interconnection are caused by imperfections within the fiber itself, losses due to the connection or installation of the connector, and factors that relate to the system such as matching of the cable to equipment.

4.6 Preparing a Connection and Termination

Most manufacturers of connectors and splice hardware will furnish a step-by-step procedure for installation. Figure 4-13 illustrates such a typical step-by-step procedure. The basic steps are as follows:

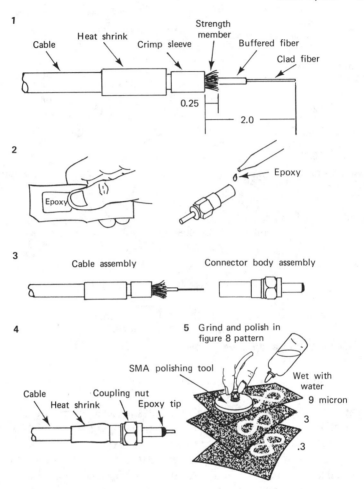

Figure 4-13 Illustration of step-by-step procedures to prepare a fiber cable for connection. Courtesy Amp Corp.

1. Cut the cable.
2. Remove the jacket, buffer tubing, and other outer coatings with cutting pliers, scissors, nail clippers, or wire strippers.
3. Remove the plastic buffer coating mechanically or chemically (special chemicals are available). Mechanically the coating is removed with a precision type of wire stripper (Figure 4-14). The stripper must be set so as not to damage the cladding.
4. Insert the sleeves in the cable.
5. Glue the connector into place.

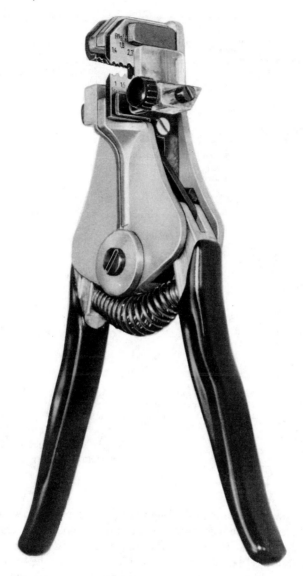

Figure 4-14 A wire stripper.

6. Heat shrink the flexible sleeve.
7. Crimp the metallic sleeve.
8. Place the connector into the appropriate holder (polishing tool) and polish the end of the fiber cable.
9. Test the cable for light transmission and if possible for attenuation.

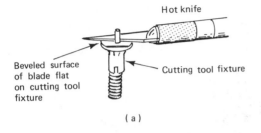

(a)

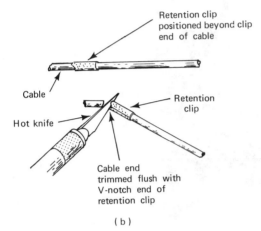

(b)

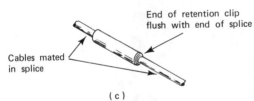

(c)

Figure 4-15 Preparing a fiber cable connection or termination in a connector.

   Plastic cable is prepared in the same manner, except the fiber cable must be cut with a sharp, hot knife to assure a flat end for polishing. This process is shown in Figure 4-15. Some success has been experienced by allowing a lens to form at the end of the fiber due to the flow of the acrylic as the hot knife cuts the material. The lens allows a good transfer of light energy without polishing the end of the fiber. Information concerning this technique can be obtained from the ESKA division of Mitsubishi Electronics America, Inc. The author's students have tried the lens technique with varied results. The temperature of the cutting blade and the pressure of the blade on the plastic fiber seem to be critical in the development of a lens that is clear and that will fit the end of the connector. The temperature of the blade determines the flow of the plastic, and the pressure determines the time duration that the blade is in contact with the plastic. If the temperature is too high or the time that the blade is in contact with the plastic is

1. Prepare cable ends to be spliced as shown in figure 1.

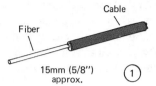

Cable
Fiber
15mm (5/8")
approx.   ①

2. Insert fiber into cutting fixture/terminal block as shown in figure 2. Cut fiber using razor blade supplied, figure 3.

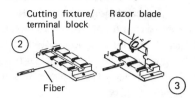

Cutting fixture/   Razor blade
terminal block
②
Fiber   ③

3. Dip the cut fiber ends into the silicone index matching gel supplied, figure 4.

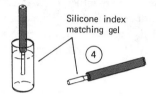

Silicone index
matching gel
④

4. Slide protective shield onto the cable and insert fiber into the splice tube, figure 5.

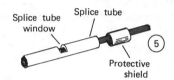

Splice tube   Splice tube
window
⑤
Protective
shield

5. Crimp the splice tube. Fill the window with index matching gel and insert the other fiber into the splice tube, figure 6. Ensure that fiber ends are in contact.

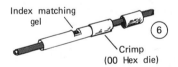

Index matching
gel
⑥
Crimp
(00 Hex die)

6. Crimp remaining splice tube end and slide the protective shield over the window. The splice is now ready for use, figure 7

Protective
shield
Crimp
⑦

Figure 4-16   Splicing a plastic fiber cable.

too long the lens will be oversized. This method seems to be a *trial and error requiring practice to make perfect.*

Figure 4-16 describes a simpler method of terminating plastic fiber. This method is recommended by the vendor for field splicing of fiber cables, for connector termination, and for repair of a break. The vendor, Nevada Western of Thomas Betts Corp., specifies a maximum of −1 dB loss for splices, and less than −3 dB loss for connection termination. Technicians performing the splices report better results of −0.1 dB splice loss and −0.3 dB connection loss.

4.7 Advantages of Fiber-Optic Cabling

Fiber-optic cabling should be considered for a new LAN system or addition to an existing system for the following reasons.

Bandwidth

The bandwidth of fiber-optic cable is over 150 Megabits per second (Mb/s) in comparison to 10 Mb/s for coaxial cable and less for twisted pair. Optical cable transmits the information by light energy instead of electrical energy and is, therefore, not subject to the electrical properties of wire such as resistance, inductance, and capacitance. These properties tend to attenuate the signal and decrease the bandpass of a system.

Signal Loss

Signal loss in optical fiber is much less than in either coaxial cable or twisted pair. Signal loss for fiber cable is less than 8 dB per kilometer. This is approximately one-tenth that of coaxial cable. This means that [to compensate for signal attentuation] an amplifier can be placed at 11 km as compared to the 1.1 km standard for coaxial cable.

Electromagnetic Immunity

The signal in fiber-optic cables is unaffected by electromagnetic noise signals. Fiber cables do not produce electromagnetic noise nor are they affected by any electromagnetic noise, such as cross talk, echoing and ringing, or static. Fiber cable operates noise free in factories, computer rooms, and other locations where electro-mechanical devices produce electromagnetic noise.

Size

Fiber-optic cable is considerably smaller than either coaxial or twisted-pair cable. In many cases fiber cables can be run under the carpet without creating a ridge in the carpet (Figure 4-7a). This can result in saving conduit that would be necessary for copper cables.

Weight

Fiber-optic cables weigh much less than either twisted-pair or coaxial cables. This can result in savings of added costs that could result from the special hangers that would be necessary for copper cables.

Safety

Fiber-optic cables carry no electrical energy so there is no possibility of an electrical spark. This allows the fiber cable to be used in explosive environments, such as chemical plants and refineries.

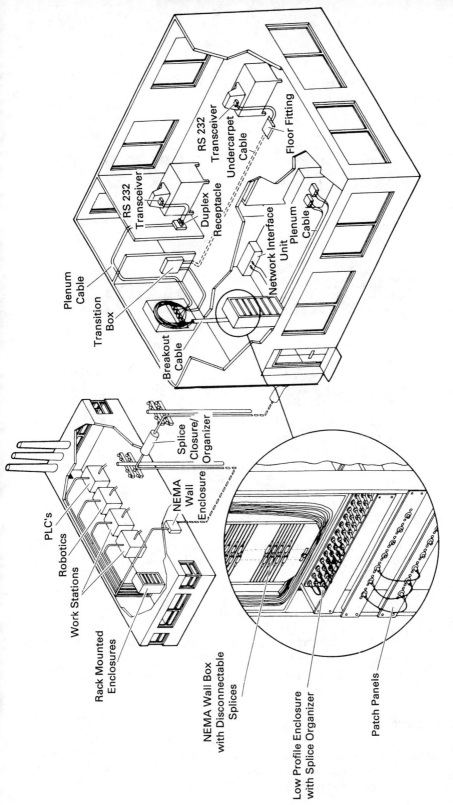

Figure 4-17 Example of a LAN interconnected with fiber optic cabling. Courtesy, AMP Corp.

NAV☆LAN ™4 Token Ring Products Application

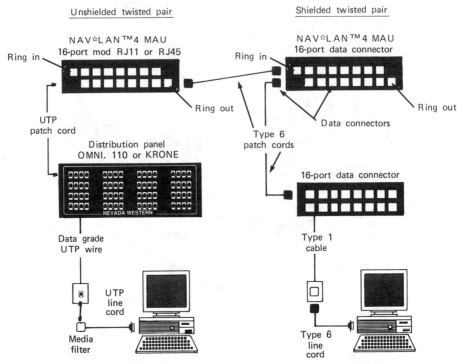

Figure 4-18 Example of a LAN cabled with a mixture of media. Courtesy, AMP Corp.

Security

Fiber-optic cables offer far more security than twisted-pair or coaxial cables. There is no electromagnetic field and it is almost impossible to tap into a fiber cable without detection.

Adaptability

Fiber cable is adaptable to almost any LAN configurations. Figure 4-17 illustrates a premise wiring implementation utilizing fiber cabling. Fiber can also be mixed with coax and twisted pair cabling (Figure 4-18). Courtesy Nevada Western

Summary

In summary, the wiring professional should balance the advantages of the three cable types against cost of installation, cost of maintenance, and future needs, before selecting a cable type. It is possible that more than one type of cable will be most cost effective for a particular campus installation. It is also possible that new technology, such as infrared transmission, laser transmission and microwave transmission, might be applied in place of some of the cabling. New technologies in wiring are rapidly forthcoming and the latest innovations should be examined before making a final plan.

Questions

1. What are the advantages of fiber-optic cable over twisted-pair and coaxial cables?
2. What are the disadvantages of fiber-optic cable as compared to twisted-pair and coaxial cable?
3. Why do fiber-optic cables offer more security of information than do copper cables?
4. What type of fiber-optic cable is most appropriate where the attached equipment must be moved?
5. What is the purpose of a strength member in a fiber-optic cable?
6. Why is fiber-optic cable the best choice of cable in an electrical noisy environment?

CHAPTER 5

Basic Network Topologies

5.1 Introduction

Network topology is the basic design or configuration of a computer network. That is, network topology describes how the various devices of the computer network are physically connected to each other. The particular type of processing needs of the business determines the type or types of network that can provide the best connectivity. Perhaps a single network will accomplish the task, or a combination of networks may be required.

5.2 Basic Network Models

There are two basic types of network models that describe most networks in use today. These are **point-to-point** and **multipoint systems** shown in Figure 5-1. Recently more complex versions of multipoint topology have been introduced to satisfy **local area networks (LAN)**. Figure 5-2 depicts simplified LAN networks. **Network topology** refers to the way in which multiple devices are connected. These connections between devices are called **communication links.** They can be phone lines, private lines, satellite channels, and so on.

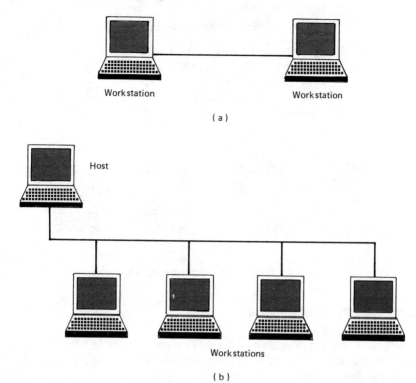

Figure 5-1   Two basic network models:
a. point to point
b. multipoint.

5.3 Determining Network Connections

Network topology is determined by the interconnection problem. For example, we have a group of devices that need to talk to each other, we will refer to them as **stations.** The stations may be computers, terminals, modems, printers, or other devices. Each station attaches to a network **node.** We must determine the type of topology that could best interconnect the stations. One solution would be to connect each station directly to every other station using point-to-point links. Obviously there are some major problems with this type of approach and perhaps some impossibilities as well. A better solution to the problem of interconnection is the use of a **communication network** that is capable of transferring data between stations (Figure 5-3).

A communication network provides a resource providing a transmission facility between stations. Several types of networks have been developed to meet different needs and all have advantages and disadvantages. We will consider these networks and their advantages and disadvantages in this chapter.

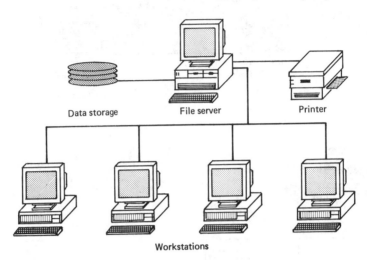

Data storage   File server   Printer

Workstations

Figure 5-2   A simplified local area network (LAN).

5.4 Point-to-Point Topologies

At the basic level, data communication takes place between two devices that are connected physically or logically together. An adolescent example of a physical or direct connection is two cans that are connected together via a string. Once the string is taut, a child at one end holds a can to his or her ear, thus hearing what is being said at the other end. For many, this may have been their first experience of *point-to-point transmission*. An example of point-to-point communication in computer systems was shown in Figure 5-1. Point-to-point nodes only communicate to others via directly connected nodes. Sometimes the nodes are physically located next to each other, but this is not necessary. The two nodes could be located some distance from each other, perhaps in different rooms.

5.5 Multipoint or Multidrop Networks

A multipoint network (some examples are shown in Figure 5-2) is one where multiple nodes share time on a line. This is called **time sharing.** Originally, this type of network was designed for high-speed data transmission rates or where there was large volume of communication traffic over the network. However, there are now several types, including the bus network. Many companies use these systems to automate their production facilities because it can be installed quickly and is very cost effective. Typical networks are the **bus network** (Figure 5-4), the **star network,** (Figure 5-5), and the **hierarchical network** (Figure 5-6). Each of these three networks will be explained.

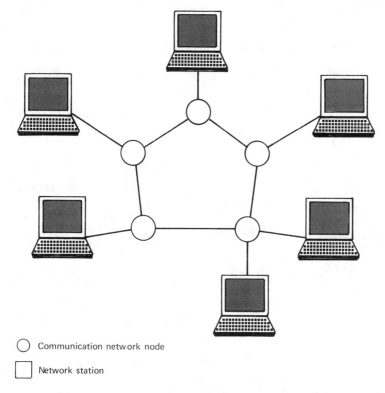

○ Communication network node

☐ Network station

Figure 5-3 An example of a communication network.

5.6 Bus Network

A bus network (Figure 5-4) has a common communications medium to which multiple nodes are attached. The bus network is a special case of a tree topology (Figure 5-7) that is characterized by having only one trunk and no branches. The tree can have multiple nodes which could be a PC's terminals, a printer, and so forth. Each component must have its own interface device. This interface device is usually a card, software, and hardware to access the network. Since multiple devices share a single data path, there must be a mechanism to determine which node will obtain the right to transmit first. This mechanism is called the **access method** and will be discussed later in the chapter.

Since the nodes share a single data path or bus, each device must have a unique address. The data are sent along the bus until the address of a specific device is found. Bus implies high speed, and bus networks are usually implemented in situations where the distance between all nodes is limited, that is, a building or a department located in the same building. The loss of a single node on a bus has little or no impact on the other nodes unless the entire bus fails.

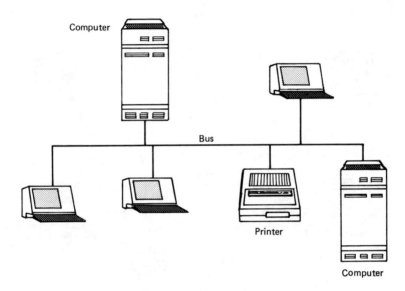

Figure 5-4   An example of a bus network.

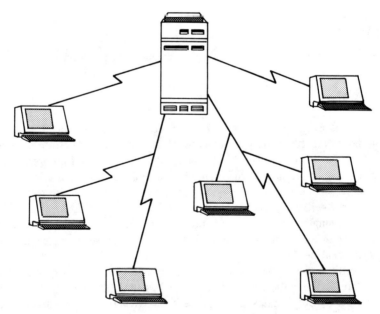

Figure 5-5   An example of a star network.

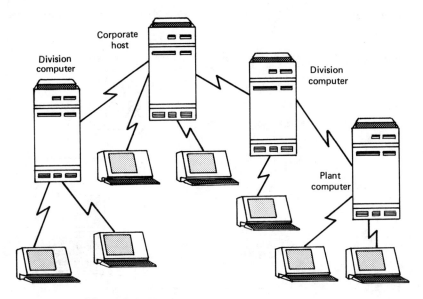

Figure 5-6   An example of a hierarchical network.

Therefore, reliability of this type of network is excellent. Each device is connected to the bus by a **tap connection** that breaks into the bus cable. However, tap connections cause a certain amount of signal power loss on the cable. Therefore, only a limited number of devices can be connected to them.

Another disadvantage with bus-type connections is that of problem isolation. Since all devices are serially connected to the bus, each device must usually be tested in sequence to locate a problem. A typical implementation of this type of network is on local area networks.

5.7 Star Topology

In the star network that was shown in Figure 5-5, a central switching network (typical of a host computer) is used to connect all the nodes in the network. The circuit can be point-to-point, multipoint, or a combination of one or more of these topologies. Such network configurations are called **hybrid networks.** Examples of these types of networks are shown in Figure 5-8. This arrangement makes it easier to manage and control the network than with some other configurations. The disadvantage of this type of network is that since all data flows through the central node, a failure causes the entire network to go down. Another potential problem with this design is that at peak periods, the central node may become overloaded and unable to transmit.

There is no limit to how many arms can be added to the star network, nor is there a maximum length of these arms. Therefore, it is easy to expand a star

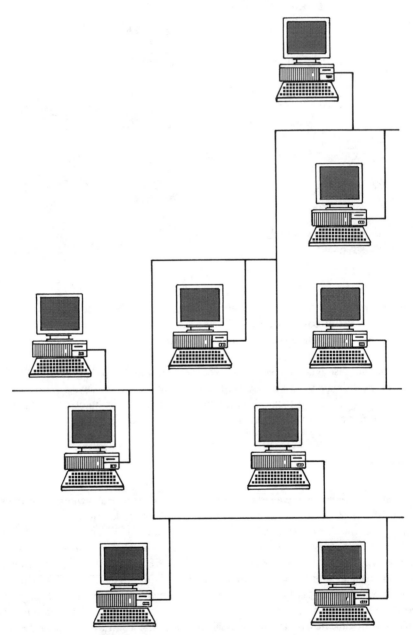

Figure 5-7   An example of a tree topology.

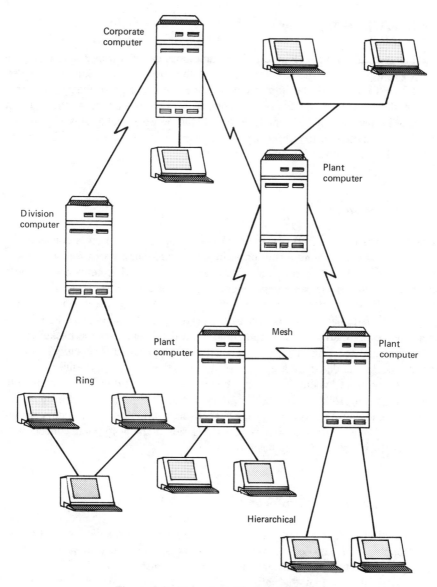

Corporate computer

Plant computer

Division computer

Plant computer

Mesh

Plant computer

Ring

Hierarchical

Figure 5-8   Examples of hybrid networks.

network by adding more nodes. When a central computer supports many terminals, the star configuration is easy to implement. Since the central computer is the master node and controls the network, the rules for the network operation are relatively simple. A typical application of the star network is the dial-up telephone system, where the individual telephones are nodes and the **public branch exchange (PBX)** acts as the central controller.

5.8 Hierarchical Topology

The hierarchical network (Figure 5-6) is sometimes called a tree structure (Figure 5-7). The top node of the structure is called the "root node." This type of network would most likely be implemented where the lower-level nodes at the second or third level are in themselves computers. One advantage of this type of network is that even if the root node failed, the network would stay up because there are computers at the lower hierarchy. Many of the features in the bus topology are shared with the hierarchical network.

5.9 Ring Network

Ring networks (Figure 5-9) consist of a closed loop. Each station is connected to two other stations forming a ring or a circle. All communications follow a clockwise or counterclockwise rotation. Most stations in most ring networks are close together, usually in the same department or room of a building.

The information that is passed along the ring is in the form of a data **packet.** The data packet that is sent by the originating station **(source node)** contains source destination **(destination node)** and data fields. As the packet is circulated around the ring, a receiver/driver (sometimes called a **repeater**) in each device checks the destination address of the incoming packet and either routes it to the station itself by simply copying the data into as a local buffer or sends in to the next station. This regeneration is important to note because it eliminates the typical attenuation that occurs during signal propagation and that occurs in bus networks. A classic ring design is really a series of point-to-point connections. However, there are many ring designs.

5.10 Network Access Protocols

We have discussed, to a limited degree, the architecture of several LANs. Each of these topologies has one common problem—that is, computer terminals, printers, application systems, and software manufactured by different vendors.

The reality is that the software of one vendor will usually not work on the PC or the workstation of the other vendor and the LAN software is usually vendor specific. In fact, some LANs have special cabling and connectivity needs.

To provide some uniformity, network user organizations have established a series of network protocols. The largest of these is the **International Standards Organization (ISO),** which has established what is called the **open system interconnection (OSI)** standards (Table 5-1).

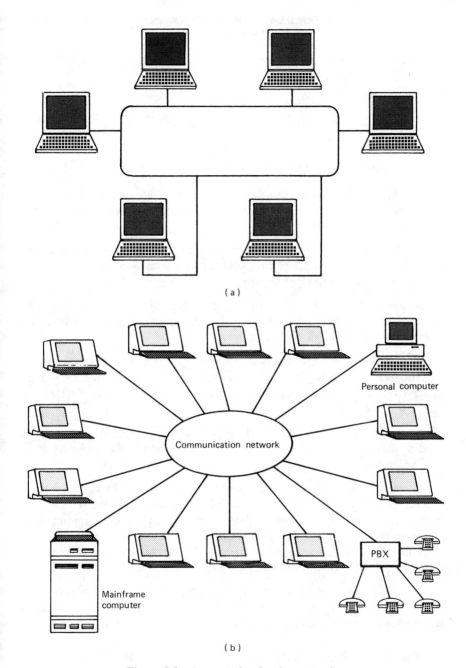

(a)

(b)

Figure 5-9   An example of a ring network.

TABLE 5-1 Levels of OSI Protocol

- Application
- Presentation
- Session
- Transport
- Network
- Data link
- Physical

The functions of the OSI levels are

1. *physical level*—a set of rules regarding hardware and transporting data across a medium. It monitors the transmission of electrical signals between two data communication stations.

2. *data link level*—this level is involved in transmitting data frames that indicate beginning and ending of a data frame. This layer also shields higher levels from the physical transmission media.

3. *network level*—this layer routes data from one node to another. This layer takes data from the fourth level and assembles it into data packets which are sent to the two lower levels for transmission.

4. *transport level*—the transport level has the functions of recognizing and recovering errors, selecting the class of service, monitoring transmission to measure service quality, multiplexing messages into one circuit, and then directing the message to the correct circuit.

5. *session level*—the function of the session level is to control when users can send or receive data, verify passwords, and determine who transmits or receives and for how long.

6. *presentation level*—the presentation level is concerned with presentation of information in a form that is meaningful to the user, network security, file transfers, and formatting functions. This layer may include code transmission, data conversion, data compression and expansion, and word processing.
This layer establishes a table of transmission code that allows one terminal protocol to address terminals that utilize a different format.

7. *application level*—performs functions that are user specific such as establish and terminate connections between users and mainframe, electronic mail, file-server and printer programs, database management programs, and so on.

The SNA Network Protocol

This model was established by IBM before the advent of the PC and is a mainframe-centered network protocol. The layers of the SNA and OSI models are similar but with differences in the service organization structures. The two architectures are not directly compatible.

Ethernet Data Packet

The Ethernet protocol was developed by DEC and adopted by several computer manufacturers and many independent vendors, and as such became an industry standard.

The IEEE 802.3 standard deals with the first two levels of Ethernet and OSI protocol. These first three levels are:

1. *physical level*—has reference on the type of transmission media: twisted pair, coax, or fiber optics; the type of transmission; the data route; and the encoding of the data.
2. *media access level*—controls how the transmission media is controlled.
3. *logical level*—allows the levels above to access services of the LAN without regard to how the network is implemented.

IEEE Network Standards

IEEE 802 was an attempt to establish standards for LANs. When the project was entered, the committee was faced with the facts that there were several established LAN standards. The architecture developed out of the committee was an attempt at flexible guidelines which could encompass SNA, Ethernet, CSMA/CD, et cetera within the OSI model.

IEEE 802 addresses only the lower two levels of the OSI, allowing the individual LAN designs or user to establish the compatibility of the upper levels.

Summary

Local area networks range from the very simple to very complex. To select the correct one for the corporation, the TCM must be well informed or an outside consultant should be employed.

Questions

1. Why is it important to establish standards for LANs?
2. What type of LAN is the Ethernet system?
3. Why did the IEEE 802 standards address only the lower levels of the IOS protocol?

CHAPTER 6

Planning the Wiring Installation

6.1 Introduction

Planning a wiring system is the responsibility of the person designated as the telecommunication manager, communication manager, or data communication manager. In this chapter we will refer to that person as the TCM. The responsibility of the TCM with respect to the wiring system is to assemble the team that will design, evaluate, install, test, and maintain the wiring system that will be utilized in the telecommunication network of an organization. The team should consist of everyone who will be involved in the project. The list may include any or all of the following, some of whom may only supply information:

- The **architect** who designed the original building or who will design the new building, if the installation is a new construction.
- The **managers** of the departments that will be using the network. They must identify the number of telephones, PCs, printers, plotters, and so on that are needed in the final installation. If the installation is to connect to an old system already in place, they must identify the types and number of presently operating equipment. It is also very important that each manager identify future needs.
- A **data communication engineer** who will evaluate the final inventory of equipment and design a local area network (LAN) that will operate the final

system. It is this person's job to select an operating system, the interfacing equipment, and the software to control the LAN.

- The **facilities engineer** who will determine the locations of equipment rooms and their suitability, the location and size of the equipment closets, the location of existing cables and if they can be utilized for all or part of the installation. The FE must also determine the location of all cabling ducts, open conduits, tunnels, raised flooring, and any other areas that might be utilized for cable installation.

- **Technical support personnel** that will be performing the wiring installation, system testing, and/or maintenance. If these people do not now exist within the company, someone must be hired or suitable personnel must be sent to school to train on these subjects.

- **Outside consultant** or **vendors** for telephone and/or data communication. Contractors can be hired for any part of the LAN installation, including system design, wiring layout, wire installation, wire testing, equipment installation, equipment testing, and system testing. The maintenance of the system can also be contracted to an outside firm.

The TCM manager should become as knowledgeable as possible about the new network and every step of the design, installation, and testing before the team is assembled for the first meeting. This is most important if some or all of the project is to be accomplished by outside contractors.

As we noted earlier in the text, it is important that planning and installation guidelines be established and adhered to whenever any major system needs to be installed, upgraded, or removed. Such guidelines should include cabling, services, and selection and installation of devices. The following paragraphs present some of the most important aspects that must be addressed and the reasons behind them. A checklist of the items should be developed and followed.

6.2 Project Scope

The scope of the cabling project should be determined before the project begins. If the project team is fortunate enough to be dealing with new construction the work will not be complicated by in-place cable and/or wiring systems. However, the following items should be considered and the guidelines should be followed for a new project, modification to an existing system, or addition to an existing system.

6.3 Existing Cabling

All existing cable and wire runs must be well documented in any cabling system. If the project requires a new cable installation in a facility where there are existing cable and wire systems, the existing systems must be well documented in advance of any work. Such documentation should adhere to the company's cable tracking system guidelines. Plans should be available showing all cable ways, under and over cable troughs, risers, and so forth, *including all unauthorized but existing cable runs.*

If the building has more than one floor, a layout of each "service floor area" is required. Each plan should include power, plumbing, and HVAC (heating, venting, air-conditioning) systems. Total square footage should be on each plan per floor or area.

6.4 User Population

The projected number of users within the facility must be known in order to estimate the number of service ports and the type of service levels that will be required from this community. This step will require input from the user community and building and/or planning departments.

6.5 Number and Type of Work Areas

The current number and type of existing work areas must be documented. This should include the number of workplaces, office spaces, classrooms, data centers, lab areas, test and engineering areas, and support areas.

It is good practice to determine if the facility work areas are to undergo any structural changes such as office consolidations, work space doubling, and/or work area conversions. Any changes will affect four areas:

1. the potential user population
2. the number and type of work areas
3. the number of service levels required
4. the type of system support required for each work space

6.6 Documentation and Room Layout Database

A physical layout of each work space and/or work area should be maintained on a database. The layout should include the following:

1. inside dimensions
2. placement and size of windows, doors, pillars, and so on

3. location and type of all ac power outlets
4. location and corporate identification number of all room cables (copper media, fiber, combined cable systems)
5. location and outlet type of all telephone equipment. This should include all modular jacks, station wiring boxes, and cable system/telephone outlets.
6. location and type of all wiring or cabling system service plates.

This type of documentation is an excellent application for a CAD (computer-aided design) system. CAD is a computer graphics program on which the designer can create, modify and display engineering drawings and diagrams. Typically, such programs are menu driven and allow the designer to rotate any part of the drawing, zoom into a specific area for close detail examination, or scale back to view the entire diagram.

An example of a two-dimensional drawing of a typical office showing various characteristics of the office layout is shown in Figure 6-1.

6.7 Type and Number of Devices Required

The type and number of dumb terminals, workstations, and word processors need to be identified. Many of these devices require special port assignments, and some will need specific controller customization. With the device type, a planner can anticipate the number and type of controller ports needed by the user community.

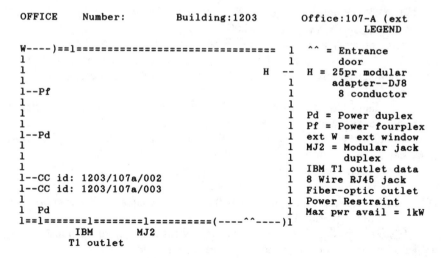

Figure 6.1   A CAD blueprint of an office.

6.8 Phones and/or Station Equipment

The type and number of phones that will be required for each work area must be determined. Presume that each area will need at least one phone if the information is unavilable at the time of inquiry.

6.9 Maximum Power Allocation and the Number of Power Outlets

Power consumption is an important issue for building managers. Even in this age of low-power consumption devices, a maximum power specification needs to be determined for each area, enforced, and periodically checked to ensure user compliance. The planning group should be able to anticipate the users' product and device mix to estimate power requirements and ensure safe power margins.

Historically, users have requested and installed more devices over time and with changing technology. As this occurs, the number of power connections can exceed the number of outlets within a given space. The users generally resort to power strips or extension cords with multiple outlet ends. This usually leads to a "power crunch." To manage ac power loading and user tendencies better, the organization should do the following in all work areas:

1. Ban the use of power strips.
2. Ban all extension cords.
3. Ban the use of plug-in multiple outlets.

If at all possible, plan to use 80% or less of power availability. This will ensure a 20% safety margin. This 20% safety margin will probably be used by the user community attaching unauthorized devices in their office or work space areas. These devices will include pencil sharpeners, answering machines, air purifiers, radios, stereos, TVs, and so on. Such noncompliance is the norm. Expect it, plan for it, budget for it.

The number of outlets will determine the maximum number of electrical/ electronic devices within each area. An insufficient number of outlets will force even the most dutiful employee to install a power connector device if the work space cannot meet minimum attachment needs.

The reason some facilities have a "power crunch" is because the initial planners never foresaw a time when users might require two or more terminals/ PCs in each office. Advance planning will ensure an adequate "power safety margin" over the long-term plan.

Historically, the first office-installed devices were dumb terminals. They needed only one ac outlet for power. However, with the advent of the PC, many things changed, including the number of individual products that required their own outlets.

In a modern office space the planners should install six to eight power outlets, due to the power requirements of current workstation systems. The following is a minimum hardware configuration for an office setup requiring a sixplex power outlet:

Outlet	Device
1.	CPU
2.	Display
3.	Printer
4.	Mouse (some require separate ac power)
5.	Fax, laser printer, expansion chassis, extra display, external modem, additional PC, desk light, radio, TV, etc.

6.10 Service Areas Affected by Work to Be Performed

It must be determined in advance which functions and areas will be affected by the cable installations. If power, telephone, or existing computer services must be interrupted in order to implement the install plan, all affected areas must be notified of the date and time of the service outage. Two weeks' advance notice should be sufficient lead time for announcements.

All pertinent information regarding service interruption and contact phone numbers must be put on all online message systems. In this way anyone logging onto a host system will get the message and know whom to contact. A contact point and meeting schedule should be set up well in advance to discuss the service interruptions with any and all interested groups.

These timely "advisory meetings," in advance of the actual work, will assist in minimizing the impact of the outages on the various departments and functions within the facility.

6.11 Review of Building Plans and Cable Requirements

The committee should review floor plan copies of *all* affected areas:

1. Telephone, voice, and data cable ways are to be documented and plans made available. These plans should include all existing cable trays, troughs, and overhead and below "raised floor" areas.

2. Pay special attention to all cable/power/coax runs that are not documented elsewhere but whose existence is known to the staff. It is important that these cable/power/coax runs are documented for future reference. It may be that they are not up to current local or corporate wiring/cabling/safety codes. If that is the case, then notify department planners to write a plan to correct any violations of current codes.

3. The location and function of all wiring closets must be documented. If your facility has dedicated communications rooms for wire/cable service as well as telephone equipment rooms, determine current service level capacities. It is important to determine the space utilization for future planning.

It must be determined if the support rooms have the floor or wall space to support more hardware. These support rooms could be corporate telco rooms (telephone equipment rooms) as well as wire/cable service rooms. Perhaps with some form of space efficiency "stacking" racks or trays, the same room will accommodate much more hardware and/or patch/LAN/concentrator racks.

4. If your facility has combined telephone and data facilities closets, and if your planning organization expects the facility to grow, then it is advisable to separate telco and data into dedicated rooms.

Note that in any facility the emphasis is usually upon the user community requirements and not support facility needs. Therefore, be forewarned, space requests for support areas may be met with substantial opposition. You may have to fight for it. Better now than later once all the space has been allocated for work areas and laboratories.

6.12 Test Equipment and Commitment to Support Personnel Training

To ensure that support personnel have the appropriate equipment to perform normal and necessary testing, installation, and repair functions, management must make a continuing commitment to training.

Training is a major issue among support departments. All personnel need to be fully trained and checked out on the necessary equipment to do their jobs. As new products are introduced into the system (be it a large data facility or a small LAN network), training must be continually updated.

6.13 Telco, Voice, and Data Support Room Requirements

The following is a partial list of some of the equipment that may be needed by the various work areas to provide adequate support services. Such areas would include all telco rooms, patch panel/distribution areas, equipment rooms, and other facility support rooms.

1. Telephone. Preferably a two-line telco circuit. In this way, a tech or support person can give assistance to the user community and still have an open line for additional support or query. It will also provide a communications link between the support room and other service centers that may be needed for problem determination and diagnosis.

2. On-line Workstation. This PC should have all levels of current and past communications programs on its hard drive. Since this is a test station, it should include all terminal emulators and control programs that are used in the facility to attach to a host. The workstation should have multisession capability. This feature will allow multiple sign-ons to assist in problem diagnosis. One session could be a host attachment, a second session would maintain contact with a service or problem history log. A third session could be used for online diagnostic support service. A fourth session could be used for online messages and so forth.

3. Network Diagnostic Tool. This could be a software package loaded into the workstation, but it might also be a separate network monitor. This will give the support personnel the ability to monitor current line condition and status. This will assist in the diagnostic efforts to solve a voice or data problem.

4. Voice and Data Test Equipment. The importance of test equipment and the proper training to use it cannot be overstated. The following is a list of some of the test equipment that might be needed to facilitate the support personnel in the maintenance of the data/voice facility.

This is a sample list, but it does give an indication of the variety and complexity of what could be needed.

1. Manual tools: universal tool kit
2. Cable and wire equipment
   Time domain reflectometer—TDR (cable break checker)
   Inductive amplifier and tone set
   Twist-on (one piece) RG coax connectors
   Crimp tools and connector parts
   Wire strippers and cutters
   Coaxial cable strippers
   Wirewrap tools (manual and electric)
   Insertion and extraction tool
   Ribbon connector crimp tool
   25-pin gender changer (to interconnect two data cables)
3. Fiber equipment
   Fiber optical power meter
   Optical continuity checker
   Fiber identifier
   Optical TDR
   Local splice alignment and measurement set (LSAM)
   Emergency optical splice kit
   Fusion splicer
4. Signal and voltage test equipment
   Continuity and circuit tester

Current meter
Digital multimeter
Oscilloscope

5. Telephone test equipment

Inductive amplifier and tone set
Punch-down tool (impact tool)
Wirewrap guns and supplies
Line test set
Modapt (eight-wire modular plug-ended testing adapter)
Modular crimp tool and modular plugs

6. Network diagnosis hardware

Breakout box (RS-232 diagnostic tool)
Bit error rate test (BERT)
Modem test set (combo break out box and bit tester for modems)
Protocol analyzer (monitor or simulation mode)
Pattern generators
Line monitor

7. Electronic repair

Solder and desolder stations
IC module insertion tool and adapters
Variable power supply
Oscilloscope
Signal generator

6.14 Environmental Concerns

All premise cable and wire products are expected to function normally in a typical facility environment. Two of these environment parameters are the temperature and humidity variation.

Premise Cable and Wire Group

For cables of all types the acceptable temperature range is $-40°$ to $+80°$ C ($-40°$ to $+176°$ F). There is no humidity specification.

Telco and Support Rooms

Due to the operating and service equipment usually maintained in these rooms, the least temperature-tolerant device will set the range for the entire room. Such "normal operating" temperature and humidity ranges will be found in the "operating specifications" portion of the user's manual for each device.

Air-Conditioning, Dust Removal, and Chilled Water Supply Requirements

Air treatment usually is required to maintain acceptable levels of temperature and humidity for the proper operation of installed machines within a computer floor or data center. The air treatment plant must be of sufficient capacity to maintain conditions within acceptable limits for current and near-future expansion plans.

Planning should incorporate forecasts of future device configurations and work loads. A plan that is able to meet requirements for the next 2 years should be considered acceptable. A 5-year plan is better and if the budget and facilities will handle it, a 10-year plan would be outstanding.

Due to certain manufacturing applications, especially in the direct access storage device (DASD) and thin film construction areas, specifications are generated concerning the air quality in a work space. In such environments, dust and/or particulate matter in the air are monitored, and if necessary, filtering systems are employed to maintain the pollutant count in the air within certain prescribed levels.

If warranted, the lab facility may have to have separate air cleaners, scrubbers, and filters to meet required specification with a work space.

Another aspect of the facility that needs to be addressed is the chilled water supply. This reference is directed at the water requirements for large CPU systems, current and proposed.

Some CPU systems, for example, the IBM 308x and 309x systems, will need chilled water service for temperature maintenance.

Plenum and Nonplenum Cable Applications

Plenum cable is a cable that has met the low flame and low smoke characteristics as required by Underwriters' Laboratory to be used for installation in air handling plenums (ducts), without conduit in ducts, voids, and other spaces used for environmental air. These cables conform to the National Electrical Code Articles 725–2(b) and 800–3(b). Each cable is comprised of insulated copper conductors often surrounded with Teflon™ or Mylar™ to give the low-flame and low-smoke characteristics.

6.15 Grounding and Bonding

As was stated in Chapter 1, electrical grounding of equipment and devices was originally intended to be a safety measure to prevent electrical shock to personnel. However, modern grounding and bonding systems are being designed to provide a low-impedance path for noise and voltage transient protection. Such electrical distributaries could disturb signals in many communications and electronic devices.

The following is a list of some of the grounding areas to be considered:

Structural ground
Signal and data path ground
Lightning protection
Existing plumbing grounds
Grounding equipment racks
Grounding racks, panels, and cable shields
Measuring building ground potential difference and ground path resistance

6.16 Cable Network Mechanical Supports

Comprised of cabling passageways, channels, and trays distribution racks, this topic will be covered in detail in Chapter 7.

6.17 Electromagnetic Interference

Electromagnetic signals can be picked up by cables and data equipment and cause data transmission error or system failure. These signals can occur from any of the following:

1. Electrostatic discharge interference (ESD)
2. Induced electromagnetic interference (EMI)
3. Low-frequency interference (LFI)
4. Radiated frequency interference (RFI)

These factors were covered in Chapter 1, 2, 3, and 4.

6.18 The Question of User Device Ownership

A big question for each organization in this section is "Who will be responsible for device ownership?" The answer will depend upon the business position the organization takes concerning ownership of the users' devices. Before this decision is finalized, the question should be answered. Will the work space device(s) be owned and therefore tracked and maintained at the department level?

If the answer is yes, then management will be responsible for inventorying and expensing the department's work space devices. This allows each department to operate as a business unit, and manage its own costs and expenses. Business justifications will have to be made to upgrade and augment current device hardware or computing facility on a departmental as well as specific employee basis.

If the answer is no, then another department, other than the users' department, will inventory, maintain, and track all device hardware. This department, for example, may be in charge of all support hardware devices throughout the function as well as the computer center support services.

Departmental device ownership has certain clear advantages over other ownership schemes, including ease of support. However, the down side is that the user community may believe it is at a disadvantage due to "packaged product and service assignments" generally believed to meet user requirements. Some users may want to be able to choose their equipment instead of having it assigned.

In either case, an inventory must be maintained and updated on a regular basis.

6.19 Hot Host Service

Another question worth considering is whether or not the company will provide at least one hot service in each work space with a known connectivity path. In this approach the telecommunications and support departments will provide at least one "hot" or "active" line per office that will provide a standard connection for a range of currently used workstations and/or PCs. Any devices outside the specified norm will have to be called into the service desk for special port assignment or software assignments. If the facility has enough controller or multiplexer ports available this is something to consider.

The advantages of this system are

1. During initial space occupancy by users, there will already be a defined system service available in each work area.
2. During moves, adds, or changes to the user population or system services, the hot service concept will greatly decrease system service connect time.

This approach will assist the strategy, planning, and support departments in driving toward a fully complemented information age office. In this office, all hardware and software services are already in place prior to the user. The connectivity paths allow almost any host-to-user path as well as provide stand-alone workstation capabilities.

Stand-alone workstations allow the user to detach from the host to do stand-alone processing. This activity frees the host for other work. This system works well in a facility that is fully populated or there is a high turnover rate in the user community.

The disadvantages are obvious. It is a resource hog. If the facility has 200 offices, at least 200 current level services need to be provided. This application commits to service 200 ports. If, by chance, the facility is not fully populated, many ports remain hot and not connected. Security is also a problem.

6.20 Building, Office, and Device Inventory

Building, office, and device inventory, to be complete, must include the following:

1. *Building and Room Location.* Each room must be documented by corporate building, floor, and room identifier. This documentation should include the listing of the current number of power outlets in each work space on the room layout diagrams.

2. *Physical Device Hardware.* This includes the type and model of each workstation device. The following example of an IBM-only shop will illustrate some of the detail needed for the documentation. The responsible group must indicate the type of terminal in each and every office space. Examples of these are

a. *Dumb terminals*

   A terminal that does not have internal memory and therefore cannot do any editing without referencing back to the host or controller will be called a dumb terminal. Some such terminals are

   3277
   3278/3178
   3279/3179
   3290—tiger terminal

b. *Smart terminals*

   Smart terminals are those which include memory. This category will include PCs and workstations such as

   PCs
   3270-PC/G/GX
   3270-AT
   5150 mdl _____
   5160 mdl _____
   5170 mdl _____

   PCS/2s
   8530
   8550
   8560
   8570
   8580

   Printers
   Remote:
       Line address: _____   Port address: _____
       System: _____   CPU or LAN attached: _____
   Local:
       Line address: _____   Port address: _____
       System: _____   CPU or LAN attached: _____

Laser:

 Line address: _____   Port address: _____

 System: _____   CPU or LAN attached: _____

Plotters

Remote:

 Line address: _____   Port address: _____

 System: _____   CPU or LAN attached: _____

Local:

 Line address: _____   Port address: _____

 System: _____   CPU or LAN attached: _____

Fax

Remote:

 Line address: _____   Port address: _____

 System: _____   CPU or LAN attached: _____

Local:

 Line address: _____   Port address: _____

 System: _____   CPU or LAN attached: _____

For example:

3274 Model—Microcode level _____

3174 Model—Microcode level _____

3. *Configurations.* Any special PC, workstation, or device configurations should also be noted in the inventory.

4. *Communications Software and Release Level.* Current installed communications software and microcode level are to be noted. Make sure equipment and software meets current standards; to this end, a physical inventory of installed inventory is necessary, checking against current level hardware and software requirements. This will help keep the user-device community updated and avoid incompatibility problems that can develop as newer levels of microcode, terminal emulators, and/or hardware improvements are introduced into the system.

Example: SYS2 Emulator xx.1 through xx.5.

Example: Please note that xx.3 requires a patch to work on system DC12. This patch is #445 and is available from department JJ15.

5. *Room Cabling Information.* The total number of connecting media within room or office space must be noted. Special attention must be paid to how many of the office cabling and/or individual coaxes are utilized and how many are not in use. These numbers become important to office/space planners as well as department and functional groups. This information can be used to arrange the user population around specific needs and services if the data are available to everyone during the planning phase of a corporate move.

Dedicated wiring and/or presence of corporate cabling system must be noted by the specified wiring label scheme.

Finally, note which cables are connected to which devices.

6. *Telephone Services.* The number and type of phone outlets by room location and all trunk and line routing must be noted.

7. *Controller Requirements.* The controller that is required or specified for each terminal service along with the controller type, serial number, configuration, and microcode level must be noted. Also indicate where controller lines are terminated. A backup microcode diskette should be maintained at all times.

8. *CPU and/or Distributed Services Including Current Status.* CPU services that are requested or required should be examined. It may be that CPU services can be provided through front-end processors or concentrators.

Maintain a running inventory of network monitor screens or concentrator screens. These screen displays should indicate current available status. That is, there should be an information panel that indicates which services are provided to the user community and which systems and services are currently "up" and running as well as which systems are "down." For those systems listed as being "down," an estimated time to be back up should be given.

6.21 Network and Plan Documentation

Network and plan documentation requires the documentation of the logical network configuration starting at the end-user devices through all connections back to the servicing centers (this could be a CPU or a concentrator). This requires the following steps:

1. Documentation of all software layers for full implementation of end users' available services
2. Recording all available software systems (MVS, VM, remote system services, DOS, etc.)
3. Recording all subsystems (CICS, VTAM, NCP, CP)
4. Documentation of the physical network topology.

These items are accomplished by showing all box connections on a two-axis coordinate system or corporate-approved location scheme for each area on the drawing. (As a suggestion, have cards or placards made up and positioned in clearly visible locations around each floor and work area with large alphanumeric building coordinates. This can be a big help in locating something on the floor or in the area. Too often this numbering scheme is placed on equipment racks or on support posts not visible from all vantage points within the area.)

There are two methods to accomplish this task. One is to use the tried-and-true manual method, which is very labor intensive. The other is to use a CAD program. This approach is computer power intensive.

In either case the drawing should reflect all areas on the floor plan, including power, HVAC venting, work areas, patch panels, equipment rooms, and wiring/service closets. These areas should have dimensions, power, and access indicated on plan.

With the manual method all drawings, other than the floor plan, should be done on clear vinyl or plastic so it can be placed over the top of the existing floor plans. This also permits stacking two or more drawings on top of each other to show various components of the network.

With a CAD system, the rooms and work spaces are displayed two or three dimensionally. With various facility service ''layers'' superimposed upon the base drawing. These would include

1. HVAC system under and above data center floor.
2. Power, including locations of outlets, types of service, and service extensions under raised-floor areas and above-ceiling locations.
3. Cables, CPUs, DASD strings, controllers, telecommunications racks and equipment, furniture, and chilled water layout. As a final suggestion in this area, the designer can use a specific color for each group or service layer. In this way a graphic with several layers superimposed on it can be more easily understood by the viewers of the terminal or printout.
4. One of the drawings should show all cable trays, pathways, and cable channels. This drawing should include
   a. all vertical and riser cable paths.
   b. all access holes, between floor riser pipes, through fire wall holes (approved and otherwise).
   c. dimensions of all sheet metal trays and boxes.
   d. dimensions and measurements of sheet metal trays and boxes above floor and below raised floor areas.
   e. radius of all tray turns or elevations.
   f. support and suspension hardware for all sheet metal or plastic trays.
      If cable trays and cableways are being ordered for installation, try to get them large enough to accommodate a long-term plan. Factors to consider when planning are
   a. tray capacity.
   b. suspension type and weight capacity.
   c. access points.
   d. cable trays above dropped ceiling panels and cableways under raised computer floors. Once installed, these are seldom if ever enlarged.
5. One drawing should show all cable wiring routing, including voice, data, video, HDTV, and fiber optic. This drawing should show all lengths of runs, drops, and splices. The individual splices may be difficult to show, but if the cable line is important, a special ''detail'' layer can be added. Any

unauthorized cable runs or cable runs that violate city building codes or corporate guidelines should be included.

An action plan will need to be developed to correct any violations of existing codes and procedures. Be sure someone is made responsible for tracking and closing out all outstanding infractions. If the city or local building department is involved, be sure all work is appropriately signed off.

6. One drawing should be only telephone circuits.

    This plan should include

    a. bulk cable routing into telco rooms.
    b. fan-outs into "66 punch-down blocks."
    c. all PBX terminations (included in this should be all APBX and CBX terminations).

7. A localized drawing can be done showing specific cabling detail. As an example, a template can be made up to show a specific wire or cable system routing, through patch and wire closets and final terminations.

8. Each floor plan should have alignment points so that one floor plan can be superimposed over another floor of the building. This will allow a planner to work effectively with an isometric building or floor plan for cable measurements.

9. Standard architectural and computer symbols should be utilized with all special characters noted in legend of plan.

10. Accurate network documentation is vital for both planning purposes as well as problem diagnosis. With a complete guide to the facility cabling system, a better understanding can be achieved and more efficient work can be accomplished.

6.22 Quality and Electronic Control

Ensure that someone verifies all cable runs for actual distance from beginning to end. This should be done for all existing cable runs to get a "cable distance baseline" for each facility.

Each type of cabling has a maximum length standard (with or without repeater or relays) as set by the manufacturer. If the cabling is part of a system, check with the system or design specifications index for maximum length statistics. Ensure that all length, splice, and signal strength criteria are met.

6.23 Service Impact Severity Classifications

A rating of "impact severity levels by work activity" must be established. These ratings must be included in all advance documentation and meetings concerning the work and any outages because of it. All department heads within the facility must be copied on all correspondence of the upcoming meetings. Include in the correspondence the schedule of meeting times and location as well as a contact name and number within your organization for additional support. Post this notice within each facility on a general news bulletin board at least two weeks in advance of the commencement of the actual work.

Finally, the "company contact" for contractors and vendors that are to be working within the facility must be notified of the "advisory meetings" date, time, and location.

All work that affects any system or service needs to be categorized by impact severity classification. This classification needs to be determined by the work or task planner. Be sure to include the category of impact in all correspondence to all department heads. It should also be a bullet item on all foil presentations. Impact categories are

1. —No risk. No impact upon services or systems.

2. —Low risk. No noticeable impact upon normal services.

3. —Moderate risk. Small chance of service hits occurring. Minimum exposure that can be handled by the installing and support crew.

4. —High risk. Good probability there will be some service interruptions due to work activity scheduled. Support staffing required. This level of impact requires one or two levels of management sign-off.

5. —Very high risk. Definite service impact due to nature of work activity. Additional field and support personnel required for installation and system restorage. Will require three levels of management sign-off and facility management sign-off.

6.24 Plan Review by All Affected Parties

All plans must be reviewed by building and area representatives, and in contact meetings, the purpose of the work must be specified. Affected departments are to be notified of actual work time, including prep time and system checkout.

Impact severity should be noted and understood by all departments. If the work requires system or service interruptions, make sure all department heads are notified well in advance so they can send a representative. At first glance this might not seem important, but departments will have personnel working at odd hours to take advantage of systems that are normally engaged during prime shifts.

Some work activities cannot be done during normal working hours, and for some departments, having personnel come in to start a backup or to rerun a big job is not that unusual.

A plan should be written and implemented to check out work when the activity is finished. This plan should provide for failure point checks and analysis if failures occur.

Backup and technical support personnel should be in place during and after work activity to ensure that a smooth systemwide transition is made from beginning of the work assignment to return of the system.

If there will be a service interruption, by doing some of the prep work ahead of scheduled "down time," the time impact can be minimized. Whenever possible, schedule all work (or as much as possible) on off–prime shift hours to minimize impact of service interruption. Perhaps more than one work activity can be combined during "normal service windows." This may take the combined efforts of more than one department, so make sure they can commit the support or work detail.

6.25 Establish a Service Desk

Instruct the work teams to follow company cabling guidelines and perform frequent inspections. Once installation teams are aware of frequent inspections, the correct labeling practice will become part of the standard operating procedure.

Establish a "service and information desk" to assist with a smooth implementation of all work activities and plans as well as serve as the first contact point for problem determination. Work procedures as well as problem determination procedures should be documented as understood by personnel who will have this job. The services provided should include guidance, consulting, and diagnosis.

The service desk work assignment will need to be filed by a person who is trained and familiar with all aspects of the facility. This should include system connectivity and user service as well as corporate long-term goals and service-level commitments.

To assist the service desk personnel, a tracking facility (file system or database management system) should be established that is available during normal working hours. Everyone in the service facility should be checked out on its proper use.

Service desk job requirements should include a technical background and diagnostic skills. The individual who maintains the service desk should be able to do the following:

1. Derive from caller all pertinent information concerning the actual problem.
2. Call up system support facilities to determine if there is an actual problem or if it is a user problem.
3. Begin first-level diagnostics once it is determined that an actual problem exists. This will determine if the problem exists in the users' equipment, line, or cable or system service.
4. Initiate, if the problem area can be identified, a problem report, send it to the appropriate department or organization, and track it.
5. Establish a number priority classification system to categorize incoming problem calls. As an example, the problem classes could range from 1 to 10

(with 1 being the least severe and 10 being a major system crash with wide-spread service interruption throughout the organization).

If, for example, a user calls up and indicates her workstation is down and a bit of checking shows the system up and other personnel within the same department are up and running, then the problem class would be a 1.

6. Determine a time-to-repair period. The telecommunications or service department's repair time by problem class should be discussed with all appropriate support service groups and agreed upon. Within the organizations, procedures will be developed to handle various types of problem calls.

7. Develop and adhere to an escalation procedure concerning the repair time window. If for some reason, the time to repair exceeds the prescribed amount for a certain level of problem, then the next level on the escalation process should be invoked. That may be additional support, head count, and/or management being brought into the picture. All repair activities should be monitored by the service desk. Each level represents support personnel with greater knowledge and technical expertise in the area. Management should be notified once a service call goes beyond the first service level.

8. Ensure all parts are ordered.
This will include required parts, spare parts, and cabling systems.

9. Establish a quality control for all cabling runs, service terminations, and device-to-service attachments.

10. Ensure that each member of the install, repair, and update team has had sufficient training to perform all aspects of work.
This requirement includes all needed parts, testing, and installation equipment as well.

11. Establish overall target dates for each project. Break down each project into subsections. Set time frames for each work subsection. Schedule status meetings prior to subsection work completion. At these meetings discuss current status of work and problems, if any, with assigned activities or support services. Review with all work groups, corporate- and vendor-level progress and overall project completion targets.

12. Maintain a clear management path for status and problem reporting by establishing a three- or four-level scenario concerning overall project completion. Anticipate problem areas and bottlenecks. Try to resolve these issues beforehand. Discuss impact levels on related systems and tasks from the best scenario of work completed on time to the worst case scenario.

13. Maintain constant contact at the service desk whose personnel will be responsible for updating the tracking facility with the work status. The tracking facility should have the capability to issue reports on current work status. These reports will include such items as subsection work completed, estimated completion time by subsection work assignments, work delays, and overall work progress.

6.26 Scheduling the Job

The TCM and his or her staff, after considering the aforementioned details, must develop a plan that will schedule every step of the design and installation of the system. This includes

- communicating with every one involved in the project, both company personnel and outside consultants, contractors, and material and equipment suppliers.
- establishing the installation schedule with specific benchmarks or steps showing progress of the project.
- coordinating contractors and vendors into the installation schedule.
- scheduling equipment and system performance evaluation test periods.
- determining the ramifications on the day-to-day operation of each department that is to be affected in an occupied area and scheduling installation work for the least disruptive times. Considerations must be given to moving a group to temporary quarters during installation.

Where installation can be performed within the work environment, consideration must be given to the effect that the installation will have on the utilization of existing equipment. When an in-place system is to be upgraded, the primary consideration is "down time" and productivity of the unit. Careful evaluation must be made of the cost of "down time" and the installation labor cost of performing the installation after hours and on weekends.

6.27 Writing the Request for Bid Proposal

The procedure for requesting bid proposals (RFP) is the responsibility of the TCM, but may be assigned to a knowledgeable person. The task must never be requested of a vendor or contractor who is to respond to the request.

The proposal request must contain the exact specification of the work to be performed and the equipment that is required. The request for proposal should be clearly written to avoid ambiguity and vague requirements. The writing must be clear and concise so that there is no room for misinterpretation by any vendor. This assures that each vendor is bidding on the task. Any ambiguities in the document allow the vendor's personal interpretation of what is to be done and may generate different responses.

Prior to preparing the RFP, the writer must know exactly what it is that is being requested. This will usually result in a great deal of research time. An RFP should never be prepared under the pressure of a deadline. Whenever specific products or equipment are required, the author must specify the manufacturer and type. The important thing is to be specific. For example,

1. To install 20 four-conductor twisted-pair 22 gauge, plenum-rated cables in the locations shown on the attached blueprint diagram.

2. To install 5 four-conductor stranded 24 gauge fire alarm and tray cables in the locations shown in the attached blueprint.

Most contractors will require a walk through the buildings in which the cabling is to be installed before making a bid. The proposal request should specify if the bid is to be lot or unit pricing. Unit pricing should usually be requested to show the cost of materials and labor for each item. This way, additions made to the bid can be priced back to the unit price.

6.28 Documentation Responsibility

A single person should be given responsibility for system and wiring documentation. Documentation should be maintained on blueprint drawings, cable distribution logs, and key sheets. The **blueprint** is a detailed drawing of all the building space. In large facilities several blueprints may be required to represent the area. The **distribution log** contains a listing of each circuit in numerical order for quick reference. **Key sheets** are charts that detail every identifiable section of the communication circuit for quick reference in servicing or changing the system.

6.29 Installation of the Wiring

The TCM should designate a single person to oversee the actual installation of the wiring and cabling to assure that scheduling is maintained and that no steps are omitted. This person must contact vendors at the appropriate time, assure their access to the facility, check the installation against the RFP, and sign off on the final acceptance test.

6.30 New Building Application

If you are involved in designing a new building plan that includes the wiring system with a **cabling approach,** the following checklist will assist you in this task.

Installation Guidelines

Ensure that all communication cabling meets the National Electrical Code (NEC) guidelines and all local codes (some municipalities have additional requirements). The NEC covers all natures of cabling such as plenum rated for air plenum environment, riser cables, strain relief minimums, lightning protection, proper fireball penetration routing and repair, and floor and ceiling penetration routing and repair. The latest code book should be consulted before the plans are complete. The NEC updates the code every three years, and this is one of several pieces of literature that should be in the TCM reference library.

6.31 Establishing a Labeling Scheme

Establish a **cable labeling scheme** that includes the origination and destination data on cable labels and tags. The labeling scheme must also include the labeling of conduit. These tags should be on a nonsmear surface, easily read, and waterproof. Both a wraparound and an in-line type should be used. Also label all conduits. The labeling data may include building and room numbers as well as sequence number for that particular room, lab, or work.

Identify all room, lab, or work space; equipment closets; junction boxes or blocks; distribution boxes; and wall outlets with proper labeling.

Develop and maintain a documentation describing all cable and optical fiber runs. This description should include point-to-point information, type of media, overall length of media, and passive and/or active components in the circuit run.

Document all hardware by creating a blueprint showing all patch panels, equipment racks, telecommunication closets, telco closets, distribution panels, concentrators, multiplexers, controllers, data circuit paths, and so on.

Establish and adhere to standard labeling for equipment and hardware. By having a standard labeling scheme for all cables and equipment, the technical team and engineers will be able to understand the layout and correct any errors. This will assure rapid turn around on future changes and additions.

Blueprint all overhead raceways/trays and distribution breakouts, and so on. The same procedure should be followed for all underfloor distribution systems.

Establish the correct media termination procedures for each type of media to ensure technical competence in installing connectors, splices, terminations, and so on.

Establish a cable database that the tracking department can use to establish, maintain, and update the database for use by the technical staff, management, and the user.

6.32 Database Tracking System

The development of a databased tracking system before cabling a new facility or adding to or modifying an old facility seems to have the cart before the horse. At least the task will seem to be an impediment to "getting the job done"—the job being the installation of the cabling, connection of the hardware, and quieting the complaints of the management and users. However, the person who is to be responsible for making changes in the system, connecting new and additional equipment, maintaining and troubleshooting the cabling, and answering all the management and user demands on the system should realize that these questions which are sure to come up are answered. This person should read Chapter 9 on wire and conduit labeling and Chapter 11 on databased wiring management and develop a plan that can be implemented as the conduit is installed, the wires are pulled, and the equipment is connected.

6.33 Safety

Safety, the last factor to be considered by most planning groups, spans several categories, for example, safety of the company employees, safety of the installers, safety of the plant, and safety of the equipment to be connected to the wiring. Here we will address the safety of the vendors and the equipment to be installed to the wiring system. The TCM or a designee should assure that the vendors are in compliance with corporate safety guidelines.

To protect the system users and the equipment that is being installed, the TCM must make sure that all local, federal, and IEEE codes are followed. Some, but not all, of these are covered in the chapter on installation of wiring systems. A current copy of the federal and local wiring codes should be part of the technical library and should be consulted to assure compliance with the codes.

Summary

The planning, installation, testing, troubleshooting, and documentation of a wiring system must be a team effort.

Questions

1. List the team that would comprise a planning group for a wiring installation in a soon-to-be-constructed building.
2. List the team that would be needed to provide wiring for a new department in an established facility.
3. When should outside consultants be hired to design a wiring system?
4. What is involved in the scheduling of cabling for a new office staff that is moving into an established facility?
5. Why is the writing of a request for a bid proposal important?
6. Who in your organization would be responsible for the documentation of a cabling project?
7. Identify, by example, the labeling scheme that you would use for new cabling that travels between two buildings into an equipment closet, through a cabling rack, and finally into a new office installation. Assume that, once the cable enters the building, half the cable travels through the ceiling and half travels under the floor.

CHAPTER 7

Installing the Cable

7.1 Introduction

The installation of communication wiring is seldom simple. There are many things that should be considered before any cabling is installed. Ideally, the telecommunication manager and his or her team would design the communication wiring for the entire telecommunication system in conjunction with the architect's plans before the building was constructed. Then all the conduit, cable trays, vertical risers, outlet boxes, distribution closets, and distribution racks could be built into the building plans. This will seldom be possible as most facilities will have been utilized by the company and cabling and equipment added as it was needed. New cables may have been added over older cables that were abandoned as equipment was updated and added.

A more realistic wish would be to have a plan and documentation of all cabling and hardware. Reality is that most buildings or areas will have installations that were undertaken by several different people who have been transferred or by outside contractors. Installations, in the past, were probably installed by each department on an as need basis, with little or no coordination, overall planning, or documentation.

Wiring systems grow as a function of increasing connectivity needs. It is estimated that wiring collections in some well-known companies, are only 20% utilized. It is often much easier and less expensive to install another cable than to find an open one.

The lesson to be learned from these companies is to *have someone in charge who will document all additions to a communication system.* This topic will be covered in Chapter 9.

7.2 Making the Wiring Plans

The communication manager or planner responsible for making adds, moves, and/ or changes will have to take things as they are and proceed from there. If this job falls to you, the following should be attempted:

1. Obtain blueprints, room layout diagrams, wiring diagrams, and any other drawings that are available for the facility to be wired. A drawing should be made with a location shown of all known cabling and communication equipment.
2. Locate all closets, conduits, raceways, subpanels, cellular floors, ducts, and so on.
3. Identify each data communication station and the type of equipment that is located at that station.
4. Obtain database information if that is available (Chapter 11).

Once all possible prints and information are assembled and the wiring manager or designer has determined the specific location and type of cabling to be installed, the TCM's team must assess the extent of the work, what resources will be necessary, and who is responsible for the installation. It must be determined if the "in-house" personnel have the time and expertise to perform the task, or if it will have to be contracted to an outside vendor.

The network wiring installations can be broken down into three categories:

1. Best situation-new building: The cabling system has been incorporated into the building wiring schematics and has been made a part of the electrical/ electronic wiring requirements for the contractors. The wiring and cable components for electrical, data, voice, video, and image transmission for all office and laboratory spaces have been identified.

This is the simplest circumstance that the telecommunication manager and support team will encounter. After the necessary performance test and turnover of the system, it should operate well.

2. Second best: An existing building with an existing system with upgrade and removal plans is in place. Major revamp of the existing system has been planned. Current and future needs have been addressed and incorporated into the plan. A plan for removal of old equipment and cabling has been included in the installation. Defective and discarded cabling that is not to be used in the revision should be removed from overhead trays, under raised floors, conduit, and so forth. The installation of all new wiring is calculated to supply current and future needs, and a documentation method is in place.

  3. Third and most typical: Existing cabling in existing buildings with no documentation and major upgrade plan is in place. Little is known about the actual capacity versus the used capacity of the cabling.

7.3 Cable Strategy

Media networks can be physically configured in several different ways. Each has its own advantage and disadvantage. There are three basic connection networks that are used in most facilities. These will be discussed in the following paragraphs.

7.4 Two-Point Connection Strategy

The most direct and simplest in terms of wiring, connections, and routing is the point-to-point connections (see topology, Chapter 5). In this strategy all host services (controller port addresses) are presented on hardware connectivity panels. These panels are usually centralized into a physical point for the service host.

  Room and lab cable runs are also terminated within close proximity to the controller port panels as described. With this method a physical connection can be easily made between host-side service to user-side devices.

  An efficient application of a two-point strategy could be for premise cabling for a small building or office with a low move-add-change requirement and internally housed computer services. All host services could be confined to a single-patch panel area, while all room cable runs would be terminated within the same place.

  Advantage. This cabling technique has the advantage of being the most direct interconnection between host and service users devices. It has the fewest breaks and terminations between host and users.

  Disadvantages. The disadvantages of this system are

  1. There will probably be a need to recable to meet changing needs.
  2. Maximum cable length could be exceeded, causing material and labor cost to rise.
  3. Propagation and signal level problems could occur with long cable runs.

  The decision to utilize this strategy must be considered carefully and with input to possible/probable future requirements. The best application would be a facility that has a very stable T/C environment with almost no moves, adds, or change activity.

7.5 Three-Point Connection Strategy

The three-point strategy is less direct but is more flexible for both users and service requirements. With this method, office/lab space cables are routed into designated "local cable rooms" that function as remote cable distribution points for specific room connections (Figure 7-1a).

A room/lab in a designated area will have cable and media runs terminate on patch or connection panels in these "local rooms." Cable feeder runs will be installed between the local rooms and the controller/host service patch panels located on or near the computer floor area.

Advantages. The advantages are

1. Host service will be broken into several smaller links betwen host/controller service panels and the user's office or lab.
2. Tracking diagnostics, auditing, and line-quality assurance will be made easier due to the segmenting of the cable runs.
3. Entire departments can be serviced by a centralized "local room" with "dedicated host system" connections. In certain instances there could be a security advantage to having sensitive host-to-user connections passing through one or two centralized points.
4. Individual cables runs are shorter, resulting in simpler upgrades and/or changes for multiple systems when they are available.
5. Service level upgrades can be accomplished through one common distribution point for a group of office/lab spaces.
6. With the installation of feeder cables there should be enough cabling capacity to accommodate future user connections. In this case, running new cabling will become an uncommon event.

Disadvantages. The disadvantages are

1. This system involves higher cost for equipment and installation.
2. It requires adequate space and more facilities planning.
3. It will add to the overall complexity of line connections between host and the user and will necessitate an effective tracking system utilizing a cable mangement system.

The system will usually require a media database. This will require labor power to create and maintain the database. This is another additional cost of providing this level of service.

Wiring Strategies

Examples of 3,4 & 6 point configurations

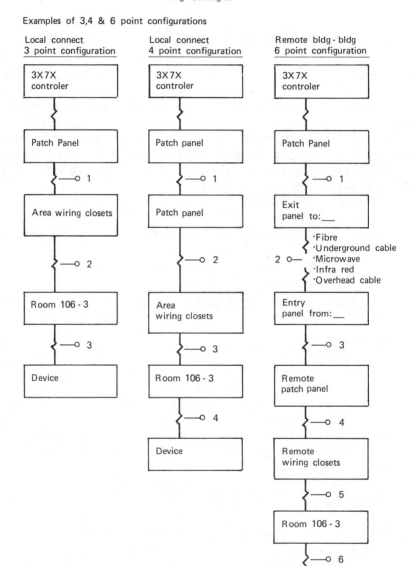

Figure 7-1   Examples of 3, 4 and 6 point wiring strategies.

7.6 Four-Point Connection Strategy

This strategy embodies parts of both the two- and three-point strategies with a twist: host services are provided within the "local rooms (Figure 7-1b)."

A small service department computer is added to the three-point connectivity link. This CPU will be servicing a specific function or task group in a local area. Its breakout or distribution panels will be set up alongside the room cable connection panels. This dedicated host will be connected point to point in the two-point connection fashion and will allow users in that area the added advantage of locally attaching to a CPU for a specific application or program. Figure 7-1 depicts examples of 3, 4 and 6 point wiring strategies.

7.7 Rules for Installing Cables

All communication cables should be installed utilizing the National Electrical Code guidelines. Communication cables are classified under the code to include fiber-optic cable, twisted-pair cable, and coaxial cable. The guidelines listed here were in effect in the 1990 code, reference Articles 200 and 800. While most of the codes are rather stable, there are changes from year to year as new products are proven or research shows that existing codes may not offer proper safety to personnel and/or equipment. The authors suggest that an up-to-date code book be part of your library.

7.7.1 Fiber-Optic Cable

Fiber-optic cables, Article 770 of the code, include

- Nonconductive cable, those that have no metal members.
- Conductive cables that contain noncarrying metal strength members.
- Hybrid cable, those that contain optical fibers and current carrying electrical conductors. These are classified as electrical cables in accordance with the type of electrical conductors in the cable.

Generally, fiber-optic cables are permitted to be included in current carrying cables containing less than 600 volts. The metallic carrying members of any fiber-optic cable should be grounded.

7.7.2 Copper Communication Cables

Copper communication cable (NEC Article 800) such as twisted-pair and coaxial cables should be supported at least 2 feet from power cables unless the power cables are enclosed in a conduit or raceway.

- **Vertical riser cable** made up of twisted pair, coaxial, or fiber shall have a fire resistance to prevent the carrying of fire from floor to floor.
- Horizontal runs shall be made so that the possible spread of fire or combustion products will not be increased by the installation of the cable. Penetration through fire walls shall be blocked.

Only fire-resistant and low-smoke-producing characteristic cables shall be permitted to be installed in ducts and plenums or other space used for environmental air. Such cabling is said to be plenum rated. Regulation information concerning cabling can be researched in Article 330-22 of the National Electric Code, entitled, "Wiring in Ducts, Plenums, and Other All-Handling Spaces."

7.7.3 Outside Cable

Communication cables are located on the same pole or run parallel to power cable:

1. The communication cable shall be located below the power cable.
2. The communication cable shall not be attached to a cross arm that carries a power cable.
3. Climbing space through the communication cable shall be provided.
4. Supply service cable having less than 750 volts running above or parallel to communication service drops shall have a minimum clearance of 12 inches at any point in the span, including the point of attachment to the building.
5. A nongrounded cable must maintain a separation of 40 inches at the pole.
6. Communication cable passing over a roof shall maintain a minimum of an 8 feet clearance above the roof of a flattop building.

The NEC gives an exception for other installations under Section 800-10 that allows communication cable to be installed in the same manner as service drop electrical power conductor (Section 230-24) at lesser heights for sloped roofs.

7.7.5 Underground Communication Cable

Again the following material is reference from the National Electric Code, Article 800.

A duct or conduit containing electrical cable shall be in a section separated by brick, concrete, or tile.

7.7.6 Communication Cable in a Conduit

Groups of cable that are run in a conduit should be pulled together. A pull line should always be pulled with a new cable to facilitate future cable installation within the conduit.

The cable manufacturer's guidelines for pulling, tension, and bend radius should be followed when a new cable is installed. The NEC guidelines should be followed for conduit size, and sufficient pull boxes should be installed in a conduit run to facilitate easy cable pulling. The conduit should probably be oversized to allow for future addition of communication equipment. The maximum total of all bends in a conduit run between pull boxes should be less than 360 degrees.

Whenever possible, twisted-pair and coaxial cable should run perpendicular to power cables. Electrical equipment such as circuit breakers, fluorescent fixtures, motors, printers, and transformers should be avoided to prevent electrical noise pickup.

7.8 Installation Techniques

The reminder of this chapter is dedicated to examples of cable installation techniques and suggestions that might be used to facilitate cabling.

7.8.1 Above-the-Ceiling Installation

Figure 7-2 illustrates installation within a relatively small area such as an office or part of a building.

In most buildings, a false ceiling is suspended from the roof in laboratory and office spaces. The ceiling panels are held in place by a lightweight aluminum grid network. This is generally done for decoration or aesthetic reasons. Between the acoustic ceiling and the roof is a "void" that is often used for facility support services. In this void space are placed air conditioning and heating sheetmetal ductwork, chilled water pipes, steam pipes, power runs, overhead cable trays, telephone cable troughs, electrical service conduits, and so forth.

As a general rule, *all cables should be run inside cable trays, troughs and/ or raceways.* However, there is an exception. If, for example, a lightweight cable is needed on an emergency or short-term basis where no cable tray or support system currently exists, it can be laid across the ceiling supports. Some sort of support should be installed as soon as possible and the cable removed from the ceiling panels. When there is anticipated growth, install a tray system or support element for the needed cable runs.

Once the needs for a temporary cable have passed, the cable should be removed for three reasons. First, the grid support system is not designed to handle much weight. It was never designed to handle the weight of cables nor the occasional tug a user may give a cable in an office space to reach a far corner in the room. Second, a temporary cable run may be strung over existing service support facilities. This can be potentially dangerous because a cable could accidentally drape itself across a connected power cord. Over time the weight of the cable could disconnect the power cord (this has actually happened).

There was another case of a cable that was thrown over a small wall (not installed inside an approved cable tray) to make a connection for a user in his

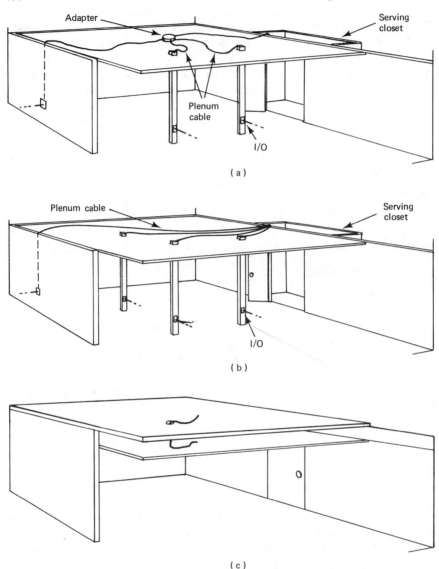

Figure 7-2 Examples of installation within a small area above the ceiling:
a. a zone-type installation
b. a home run–type of installation.

office. The cable developed a loop that got caught up on a pressure relief valve. Once the receiving end brought the cable down into the office space, it was determined that it was just a little short to get to the workstation. The user pulled the cable and the loop drew tight against a relief valve lever. Since the user thought the cable had gotten caught up on something he thought another tug wouldn't hurt. You can guess the rest.

Third, the use of temporary cabling sets a very bad precedent that office mates are sure to follow and copy. Once caught, the office worker will use the non-complying cable connection as an excuse.

Figure 7-2a illustrates a zone installation. The area is served by a cable from the closet to a multiple-cable adapter from which drop cables are connected to the individual communication equipment. Figure 7-2b is an example of an over-the-ceiling home run. Here each piece of communication equipment is connected directly to a source cable at the home closet or cabinet. Figure 7-2c represents the simplest method of adding a single piece of communication equipment. The cable is simply punched through the ceiling and connected to the communication equipment. No special fixtures, channels, or ties are used. However, whenever possible, all cables should be supported by ties to structure other than the ceiling.

The most permanent overhead installation and the most costly is to install conduit as shown in Figure 7-3.

7.8.2 Under-the-Floor Runs

If the building were designed with future cabling needs in mind there will probably be facilities for under-the-floor cabling.

Figure 7-4 illustrates an under-the-floor conduit system. It is hoped that room was left for additional cables and a pull line was left in the conduit. If these two factors are met in the construction of the facility or past wire additions, the pulling of additional cables should be simple. All the cables should be pulled together. It is advisable to use an inert "cable grease" on the cables to prevent binding with other cables in the conduit. Cable "grease" looks like liquid soap,

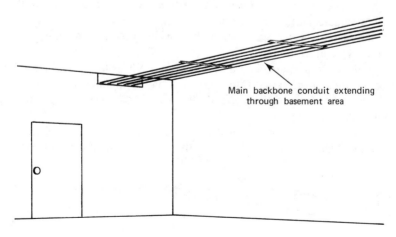

Main backbone conduit extending
through basement area

Figure 7-3   An example of a conduit installation.

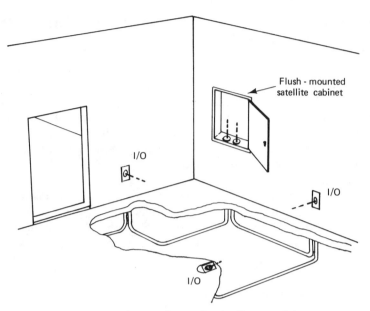

Figure 7-4   An example of an under-the-floor conduit system.

but it is a lubricant that will not harm the insulation on the cable and dry in a short period of time.

Figure 7-5 depicts a wire mesh cable grip that can be used to pull cables through conduit. The grip is available in many sizes and must be the proper size for the cable or cables to be pulled. When multiple cables are to be pulled, the total diameter of the bundle must be measured and must fall within the circumference range of the cable grip.

Some buildings were designed with removable floor sections above the concrete floor (Figure 7-6). This type of floor can be added to almost any room and may be advisable where large numbers of wires are necessary to supply the communication system in the area.

Figure 7-7 is an example of a celular floor. Cells or tubes were placed in the floor support under the raised floor; electrical and communication cables are run through the cells. Communication and electrical wires must not be run through the same cell.

Figure 7-8a illustrates under-the-floor ducting that was installed under a raised floor. Ducting was provided for both electrical and communication cables.

Wire pulling grip

Figure 7-5   An example of a cable grip used to pull cables through a conduit.

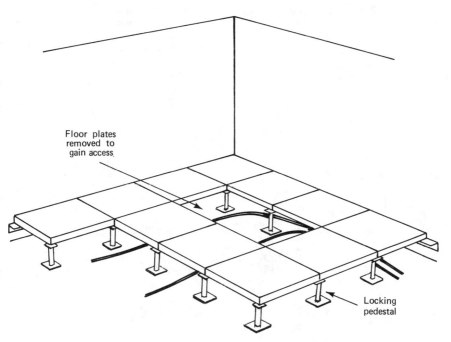

Floor plates
removed to
gain access

Locking
pedestal

Figure 7-6 An example of a raised removable floor.

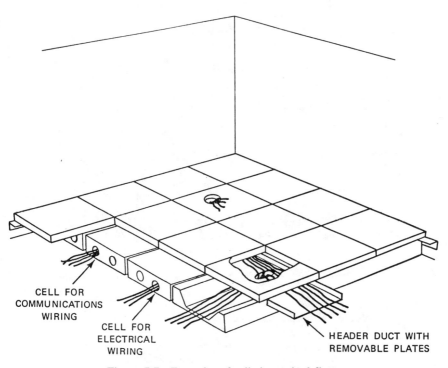

CELL FOR
COMMUNICATIONS
WIRING

CELL FOR
ELECTRICAL
WIRING

HEADER DUCT WITH
REMOVABLE PLATES

Figure 7-7 Examples of cells in a raised floor.

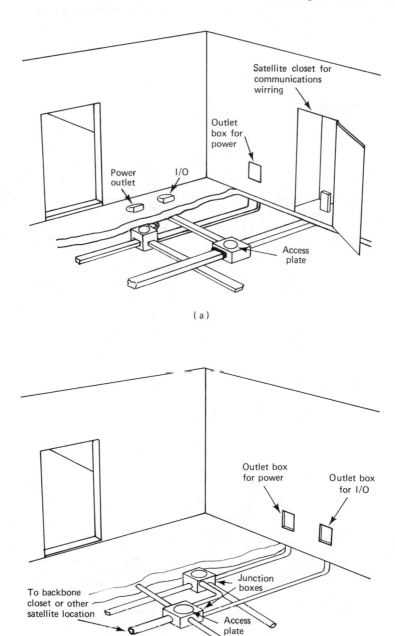

(a)

Figure 7-8 Examples of ducting in a raised floor:
a. two-layer ducting
b. single-layer ducting with junction boxes and access plates.

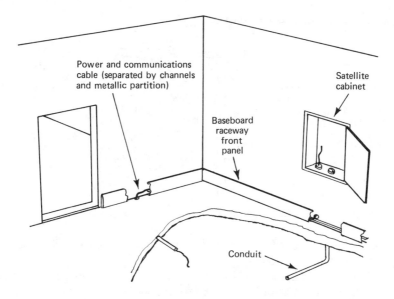

Power and communications
cable (separated by channels
and metallic partition)

Satellite
cabinet

Baseboard
raceway
front
panel

Conduit

Figure 7-9   An example of a baseboard installation.

Note that a separate closet is provided for the communication connections, and an outlet box is provided for the electrical connections to keep the two isolated.

Figure 7-8b illustrates another example of under-the-floor ducting. This ducting is on one level. This type of ducting is necessary where there is insufficient space for two levels of ducting. The ducting has junction boxes with access plates to allow for pulling or making cable connections. Copper communication cable and electrical cable should not be placed in the same duct.

An example of baseboard installation is shown in Figure 7-9. Hollow baseboard can be installed in almost any area. The cable is routed behind the baseboard to the equipment. The cable can be run into another room by making a hole in the wall under the baseboard.

Figure 7-10 represents a variation of the baseboard method by utilizing hollow molding. The molding can be installed at any height on the wall or in the corner where the wall and ceiling meet.

Vertical connections can be made by using a wall strap as shown in Figure 7-11. The cable is clamped to a steel support bracket that runs from the ceiling to the floor.

Whenever overhead cables are connected between buildings from a pole (Figure 7-12), a steel support strand must be included in the cable and secured to the pole and the building by appropriate hardware. If the cable is run down or along the building it must be protected by a metal or plastic EMT cable guard.

In some buildings tunnels will have been included for cable installation (Figure 7-13). Separation of electrical and communication cable is very important in a

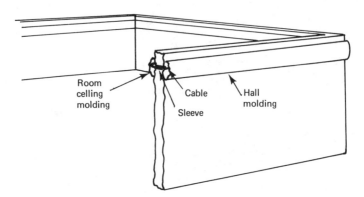

Figure 7-10   An example of a molding cable installation.

tunnel installation. The NEC must be followed and the cables must be magnetically separate to prevent electromagnetic pickup.

     Cabling can be purchased with a UF insulation rating that can be directly buried in the ground (Figure 7-14). The UF-type insulation is a tough plastic that is not affected by water or Qf soil acid. However, gophers and ground squirrels love it. The small additional cost of EMT-type conduit is an investment well made.

     Under normal conditions, typical and usual methods of cable installations, access, and routing are applicable. However, in the case of access into a controlled environment, a more restrictive set of access rules applies. For example, a research and development group laboratory or competititve-analysis room within a controlled space (such as a computer floor), would generally be classified as a *controlled area*. Corporate and site guidelines should be developed and implemented to restrict access to cabling to these types of functions by controlling ac-

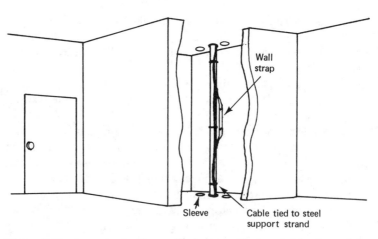

Figure 7-11   An example of a vertical riser cable run that is clamped to a steel brace.

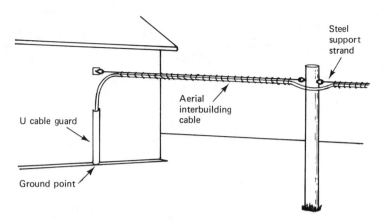

Figure 7-12   A cable drop from a pole to a building.

cess to cable trays and overhead cable support systems. Approved methods and access points must be established prior to cable routing taking place. Figure 7–15 depicts an example of a controlled access room. The security system for the area can vary from a simple lockable feeder cable opening to dedicated individual fiber and cable lines each with alarm-equipped access covers and video surveillance in place. There are three areas of exposure that need to be addressed when dealing with security or restricted access rooms:

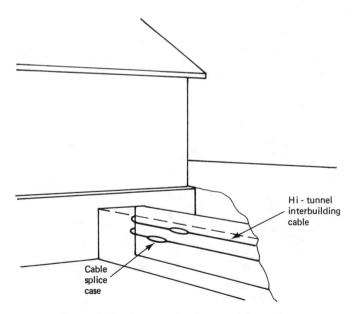

Figure 7-13   An example of a tunnel for cable.

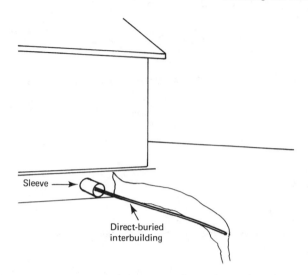

Figure 7-14   Special types of cables can be buried directly in the ground.

1) below secured room access (cable trays under a raised computer floor)
2) at secured room wall access
3) above secured room access (through overhead cable and wire trays)

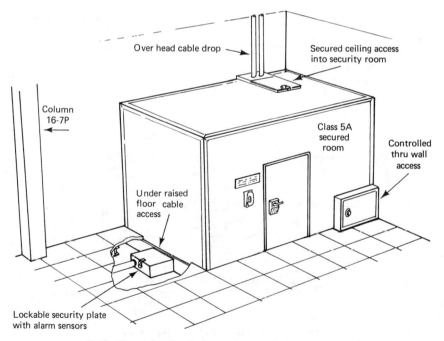

7-15   Example of a designated controlled access room.

Cable patch panels and breakout boxes may need to have some type of security system.

7.9 Cable Installation Hardware

There are many manufacturers of cabling hardware and thousands of different hardware pieces, each designed to make the job of a cable installer easier. Before attempting a cabling installation, addition, or modification of a cabling system, the installer should review several catalogs and become knowledgeable about cabling hardware. Most manufactures or distributors will furnish, on request, details on the installation of their products. In this chapter we will review only a sampling of these products as examples of cabling hardware.

7.9.1 Overhead Cabling Hardware

Most office buildings have a false ceiling installed several feet below the roof for aesthetics, acoustics, soundproofing, heating, and cooling. The "dead space" above the ceiling is used to conceal the heating and cooling ducts, electrical conduit, and water and gas lines. This space is an ideal location for most communication cables.

When an area requires only a few cables, they can probably be placed directly on the ceiling supports. However as the telecommunication system grows and the cable requirement increases, more support is necessary. Figure 7-16 illustrates a cable runway that is supported by an overhead bracket. Brackets can be added to the cable runway that enable change of elevation or direction as desired.

When an area requires a large number of communication and data line drops, a grid support structure is more suitable. The grid structure depicted in Figure 7-17 has a hatch installed for overhead access. Both the cable runway and the grid structures can be supported by threaded rods from the roof or by floor column supports.

7.9.2 Central Distribution Points

As a telecommunication network grows there is increased need for a large number of user terminals and other equipment to access the host computer. A growing network requires that centralized information access points for cabling, whether it be twisted pair, coaxial, or fiber optic be established.

The most common central distribution point is the terminal block used for twisted pair simply because it has been used for many years by telephone companies. Figure 7-18a depicts a 66 terminal connectorized punch-down block. The numbering convention for such a block is illustrated in Figure 7-18b. In a small facility a block or two may be fastened to a wooden wall by screws. However, larger facilities such as that depicted in Figure 7-19 require that methods offering easier cable identification and better tracking control be utilized.

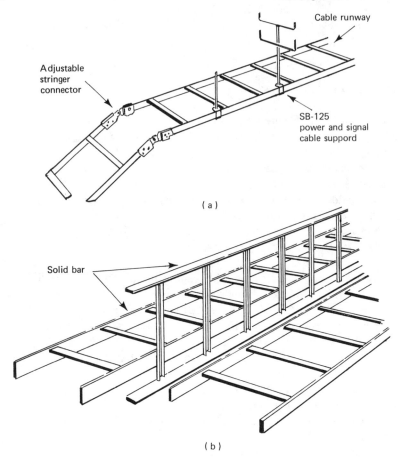

(a)

(b)

Figure 7-16   A cable runway.

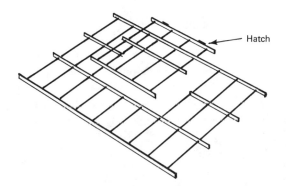

Figure 7-17   A grid-type overhead cable support with a hatch opening access.

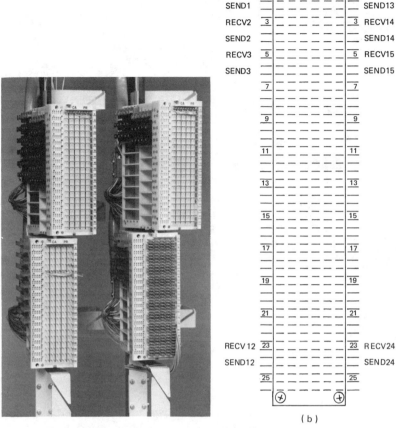

Figure 7-18 A telephone punch-down block:
a. photograph
b. a simple numbering system for the block. Courtesy, AT&T.

Figure 7-20 illustrates a typical communiction rack that is loaded with relay racks. The distribution racks should be installed in a preselected dedicated area (usually called a closet). The closet should be clean, dry, and cool, with ample space for cable installation, modification, and testing. Space should also be dedicated for future expansion of the network. A small area may "make do" with a rack cabinet.

Terminal blocks with punch-down connections are common in older facilities. However, designers of newer installations are looking to different and better methods for cable connectability. The two most often utilized connectors selected for twisted-pair cabling are the RJ11 (6-pin) and the RJ45 (8-pin) connector shown in Figure 7-21. These connectors have become standards in the modular telephone

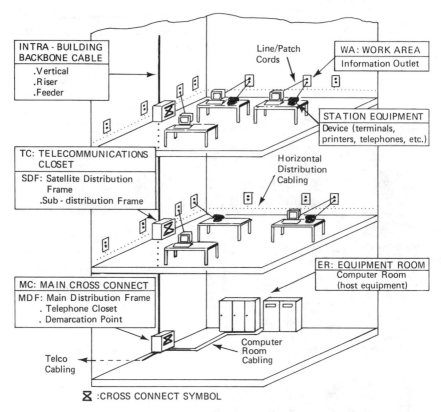

7-19 A multistory LAN.

systems and are often used as the input and outputs for PCs and other digital equipment. Their adoption was spurred when manufacturers realized that only a few of the 25 pins were being used in the EIA RS-232 DB25 connectors (Figure 7-22).

Patch panels utilizing the RJ telephone connectors allow easy cable rerouting without the need for tools.

Coaxial cables are terminated with either BNC or TNC connectors and have different characteristic impedances than twisted-pair or twin-axial cable. To prevent a mismatch of impedance and a corresponding signal attenuation, an impedance-matching device called a balun (Figure 7-23) is connected between the TWP and the coaxial cable. A connectorized balun can be connected at any point between twisted-pair and coaxial cable. The baluns can be mounted at a cable panel that can be mounted in a rack with punch-down blocks, between the workstations and the wall plates, and so on. Figure 7-24 illustrates a system utilizing twinax and coaxial cable with a balun-matching connector panel.

Coaxial cables are rather heavy and should never be hung on their terminal connections or stretched. Strain relief should always be provided to prevent the cable from separating from the connector. Cable ties should be used to fasten

Figure 7-20 A communication cable rack with terminals blocks installed. Courtesy, Nevada Western.

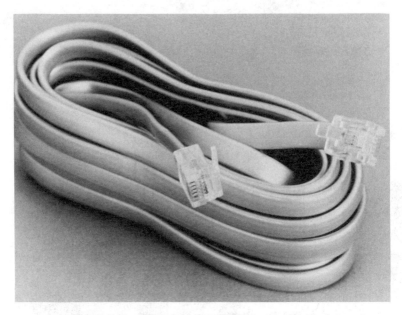

Figure 7-21 Examples of an RJ type extension cord.

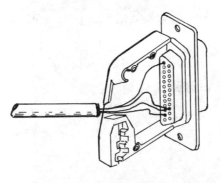

Figure 7-22 An example of an EIA RS-232 DB25 connector illustrating that most applications utilize only a few of the pins.

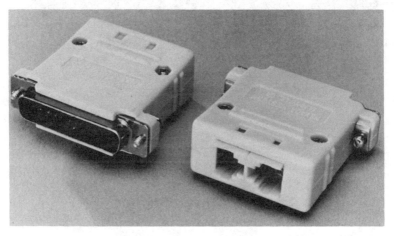

Figure 7-23   A balun impedance matching device.

coax to a solid support both in the ceiling, and especially when the cable is run up walls between floors. The ties should be tight enough to hold the cable but not so tight as to crush the cable. If the tie is too tight, the dielectric insulation will be crushed and the outer shield can short to the inner conductor. When coaxial cables are connected into a patch panel a strain loop must be provided. Figure 7-25 illustrates a patch panel cable guide across which coax can be draped.

Care must be taken when coax is laid on the ceiling support or an overhead cable grid so as not to strain the cable as it is being pulled across the overhead supports. This can be prevented by allowing cable-pulling loops as depicted in Figure 7-26.

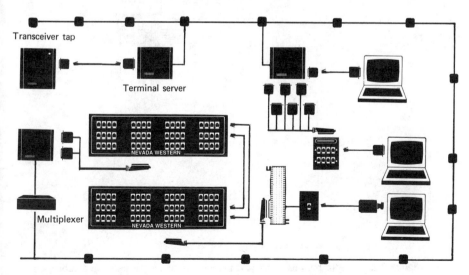

Figure 7-24   A system utilizing both twisted-pair and coaxial cables. Courtesy Amp Corp.

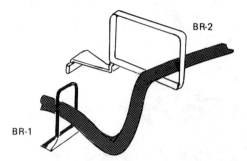

Figure 7-25 A coaxial patch panel guide. Courtesy, Automatic Tool and Connector Company.

Fiber-optic cabling requires fewer actual cable runs. However, the electrical signals at the source must be converted to light signals for the fiber cable and reconverted to electrical signals at the destination. This can be accomplished by a mux at each data processing device or at a central distribution panel. The latter is by far the more economical. Figure 7-27 illustrates a multistory-multibuilding LAN that utilizes all three communication wiring media. A single fiber-optic

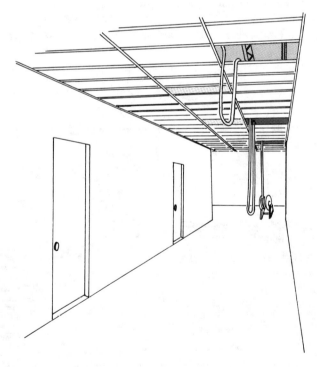

Figure 7-26 Cable pulling loops will prevent strain on the cable and cable connectors.

cable connects the buildings, and coaxial and twisted-pair cables connect the telephone closet and various racks on each floor.

7.10 Grounding the Cabling System

We have mentioned the necessity for proper grounding of equipment and cabling several times in the text; however, it is such an important point that it bears repeating here. The grounding of equipment is for the most part for the safety of the workers. An electrical short in ungrounded equipment can result in the outer case being at line voltage in respect to another piece of grounded equipment, a conduit, the earth, or a concrete floor.

As we stated earlier, electrical noise is the introduction of any interfering voltage developed internally or externally to an equipment or transmission line. The larger the number of electrical/electronic devices within a premise the greater the possibility for inducing electrical noise within the system. The most likely pickup of electrical noise would be

1. Low-frequency interference (LFI) usually 60 Hz of electromagnetic radiation from power lines, fluorescent lights, and so on.
2. Electrostatic discharge interference (EDI) forms electrostatic voltages from motion of people or rotating electrical devices.
3. Introduced electromagnetic interference (EMI) from electronic devices such as computers or other data transmission lines.
4. Radiated frequency interference (RFI) is electromagnetic interference from signals in the air such as radio signals or microwave signals.

Grounding of the plant, the equipment, and the cabling must meet the National Electric Code requirements for proper grounding. The latest edition of the NEC manual should always be consulted because the requirements are sometimes changed or modified.

We recommend reviewing Chapters 1, 2, 3, and 4 for suggestions for the proper grounding techniques for twisted-pair, coax, and fiber-optic cabling and plant grounding.

Summary

There are as many cable schemes as there are buildings. The building topology, the present inventory of data processing equipment, the current in-place wiring, and the future telecommunication plans will all dictate the type of network and wiring media employed. The wiring media and wiring methods should be tailored to the facility and the present and future needs of the business. The wiring of a network should be given as much attention as is given to the selection of the data processing equipment.

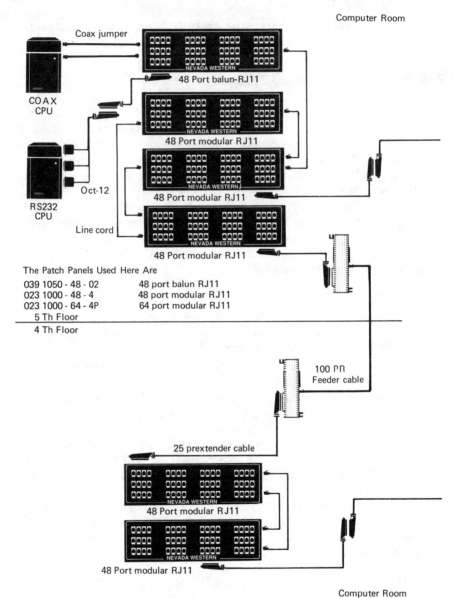

Figure 7-27 An illustration of a multistory-multibuilding wiring system for a LAN.

Questions

1. Why should all cabling be pulled in a conduit at one time?

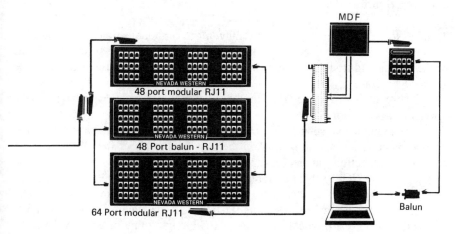

2. From the point of view of the wirer, compare the installation problems of twisted-pair, coax, and fiber-optic cabling.

3. How would you determine which cables are out of use in your facility?

4. Compare the procedure of "finding" an unused cable and adding a new cable for a user.

5. Why is the proper termination of a coax cable important?

CHAPTER 8

Testing and Troubleshooting

8.1 Introduction

Testing, troubleshooting, and maintaining a telecommunication system are involved procedures that encompass both hardware and software specialists. Some of the testing and troubleshooting can be accomplished by the use of a diagnostic software program.

These functions of testing, troubleshooting, and maintenance fall in the lower level of system protocol.

For example, personnel performing troubleshooting, maintenance, and preventive maintenance would operate in the first four levels of the OSI (open systems interconnection) model developed by the International Organization for Standardization (IOS) shown in Table 8-1.

Protocol systems assume that the various wiring media are intact and functioning correctly. Therefore, installing, testing, and maintaining wires and cables fall below the lowest protocol level, in what we might think of as the 0 layer.

8.2 Objectives of Testing and Troubleshooting

Testing procedures and methods have three objectives:

1) Train personnel to ensure all specialists, engineers and technicians who are required to perform test sequences are checked out on all test equipment.

TABLE 8-1   The Layers of the OSI Protocol

Layer
1 Physical layer
2 Data link layer
3 Network layer
4 Transport layer
5 Session control layer
6 Presentation control layer
7 Application layer

As part of this requirement, each operator should take a "hands-on" test periodically to ensure correct technical analysis is derived from test equipment data.

2) Monitor existing test and service circuits as part of an ongoing preventative maintenance schedule requiring periodic checks and measurements.

3) Locate and identify defective components within the system. This should be accomplished by systematically testing by functional groups. This procedure is sometimes called "bracket-troubleshooting" and is helpful in doing corrective maintenance.

The issues of functionality and testability need to be addressed for any equipment or cabling system that is to be installed. Functionality is self-explanatory. However it is more difficult to incorporate the second item of testability into each circuit or system.

Some designers plan in test function devices as part of the system from the beginning. Other designers provide break-out test points so that test equipment can be "hung-on" or switched into the circuit in order to test it. Poor design does not address the problem of test access at all.

There are three testing access needs to be addressed in system testing: manual access, switched access and in line access. In line access may utilize both manual and automatic testing and analysis methods.

The manual method usually employs the use of test jacks, test points, and/or breakout panels of individual components sometimes through test patch panels and terminal strips. The technician will generally need to bring the test equipment to these locations in order to make the test and determination of where the source of the problem is. Usually such equipment must be small and portable since the test points may be at different locations throughout the system. These access points allow test equipment to be hung-on to the circuits at various points to determine functionality. Test points should be placed by the cabling system designer in locations that will "bracket" or break the system into its logical components. However, sometimes when you are trouble-shooting a problem on a system with prescribed test points, you may find that you need to test a sub-section or circuitry for which no test jack was provided. This method is usually the most labor intensive and least equipment intensive since the same testing equipment can be used for various tests and different locations.

The switched-standby method requires that the test equipment be positioned near or at the circuit test points. It can be "plugged or switched" into the line or prescribed test points. This method is more equipment intensive than the manual method access but is more efficient because there is little equipment setup or installation time. Test equipment stations are installed at critical points throughout the system. Each station will have the necessary devices to perform various types of analysis based on a predetermined number of tests. Generally there will be some equipment redundancy with this method.

The in-line access (manual analysis) method uses passive devices that are actually installed in the circuit. Overall system performance can be slightly degraded by the addition of passive test equipment, but its advantages are many. The equipment can serve as a "performance monitor" of ongoing system or circuit activity. This monitoring is especially helpful when introducing new or different devices into the network or system. The introduction of new devices may cause a change of transmission characteristics that could change or degrade the signal transmissions. This method is the most cost-intensive and typically employs the most sophisticated equipment.

In-line access (automatic testing and analysis) requires that some in-line devices be used to perform *automatic line testing*. Such devices are computer based and can be programmed to run a variety of tests and functions on an ongoing basis without intervention. Once the testing phase is finished the device will generate a test data file. This file can be routed to specialists for review and analysis. If the data indicates a problem, appropriate action can be taken to verify it and, if necessary, imitate a locate and repair plan.

The data record can also be sent to an "analysis program" for review. Such programs take the raw data (usually numeric), perform calculations with the input, and compare the data against predetermined limits. The results of this analysis are then sent on for automatic distribution via a routing list. When the resultant values exceed predetermined limits, a warning or alarm can be given to notify a technical support center of the problem.

The following would appear on the monitor screen

!!!   RED FLAG ALERT   !!!

1) DOCUMENT AND CORRECT ALL INACCURATE LABELS.

2) DOCUMENT AND CORRECT ALL IMPROPER CABLE RUNS, CONNECTOR-TYPES AND/OR INSTALLATION.

3) DOCUMENT AND CORRECT PATCH PANEL, BREAKOUT BOX AND TERMINAL LABELING.

4) UPDATE DATABASE EVERY TIME A CORRECTION WAS INITIATED OR PERFORMED.

Please be sure to look for any of the above mentioned 'RED FLAG ALERTS' during your work activities. If any are spotted, correct as necessary or initiate appropriate paperwork to do so. And be sure to update the database system information.

Testing of each type of wiring medium described in Chapters 2, 3, and 4 may be somewhat different but the purpose of testing is the same in each case, that is, to provide a medium over which to transport voice or data information with a minimum amount of loss, distortion, noise pickup, and cross-talk.

We will examine methods used to test each type of wiring media: twisted pair, coaxial cable, and fiber-optic cable.

8.3 Testing Twisted-Pair Wires

Any time that cables are installed or there is a cabling change for installation for new equipment, the installation and testing of twisted-pair cables should include:

1. the correct labeling of all wires and all termination points. Wire labeling may, in small installations, be by color of the wiring insulation and the code placed on a wiring list. However, each termination point must be labeled to identify it on a wiring list or database. Labels at termination points or terminal connections should identify the difference between telephone and data lines.

2. the testing of each wire within an installation for continuity, opens, shorts between wires, shorts to ground and hi z shorts.

3. polarity testing to assure that the signals from a source will arrive at the destination with the proper polarity.

Any time there is a cabling change, entailing addition of cables for upgrading, equipment replacement, or relocation of equipment, every cable must be tested for continuity or a short circuit, and every connector must be checked to assure that it is correctly wired. This may seem like an unnecessary procedure and a complete waste of time, and it is 99% of the time. However, that 1% could cause days of troubleshooting of the system to find a short or open circuit after the system is complete. It is also important that any cable changes by entered in the database.

8.4 Continuity Test of a Cable

A continuity test assures that a cable is complete from end to end. This test is accomplished with an ohmmeter found in a volt-ohmmeter (VOM) or in a digital voltmeter (DVM). A DVM used on the "diode" scale will sound a tone when a short circuit or continuity is found. A device called a continuity tester can also be used for the test on twisted-pair cable. Continuity testers are simple ohmmeters that give an indication of a short or continuity by a tone or a light.

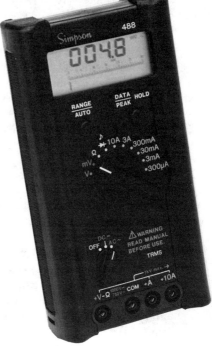

Figure 8-1 Examples of:
a. an VOM OHM meter
b. a DVM.

The VOM (Figure 8-1a) has an analog scale. When the test leads are connected together the meter pointer will read zero on the ohms scale. When the ohms function is selected on the VOM the readout will indicate zero ohms (Figure 8-2a).

Finally, when the test leads are shorted together on the continuity checker, the light will be illuminated. More expensive continuity testers will sound an audible alarm when continuity is complete.

When the VOM is used for a continuity or short test, the test leads should be shorted and the pointer adjusted to the zero position with the "ohms" adjustment. This adjustment compensates for aging of the internal battery in the VOM (Figure 8-2b).

To determine the continuity of a short cable, the VOM or DVM can be connected to a wire on one end of the cable and the other lead of the instrument connected to each wire at the other end of the cable until a zero response is indicated (Figure 8-3). When the wire is found, the pointer on the VOM will indicate zero or the readout on the DVM will indicate zero ohms. When the indicator is used, the light will indicate continuity. An open wire would result in no indication on either of the instruments.

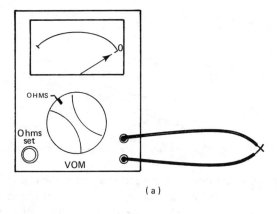

(a)

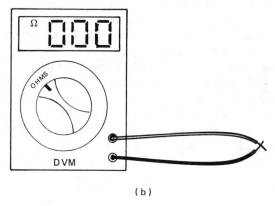

(b)

Figure 8-2   Setting of the OHMS adjust on the VOM and DVM.

8.5 A Short to Ground Test

A short to ground (grounded wire) can be found by connecting one lead of either instrument to ground and then connecting the other lead to one wire at a time. A shorted wire to ground would give the same indication on the instruments as wire continuity.

Each wire should also be tested against every other wire to assure that no two cables are shorted to each other. This is accomplished by connecting one lead of the continuity/short testing instrument to one wire and then connecting the other lead to each wire in turn. Then another wire is selected and the process is repeated until all wires are tested to every other wire. A record of each wire test should be recorded to simplify the test. For example, in an eight-wire cable, wire 1 would require 7 tests, wire 2 would require 6 tests, wire 3 would require 5 tests, and so on.

The testing of a long cable is somewhat more difficult in that it is impossible to connect the testing instrument lead to both ends of the cable. However any

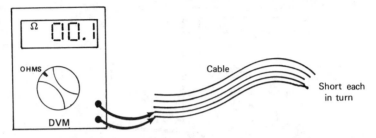

Figure 8-3   Using the ohmmeter to test for continuity on a short cable.

two wires can be clipped together at the end away from the test instrument and the resistance of both wires read at the other end of the cable (Figure 8-4). The resistance of the wires should be very low (see Table 8-1) and can be estimated if the length and wire size are known. The wires must be clipped to the first wire one at a time and tested or all wires terminated with a termination box. The resistance of any pair of wires should be low and almost the same value. If the cable is very long, two people can perform the test communicating with a two-way radio or by telephone.

A ground wire test on a long cable is performed in much the same way; one lead of the test instrument connected to a ground that is common to both ends of the cable (Figure 8-5). The wires are then connected to the continuity instrument one at a time. The other end of the wires must be open, and this test can be performed by one person. Only those wires that are purposely connected to ground should show a response.

A short between wires would be performed by one person in the same manner as was described for a short cable.

8.6 Testing a Coaxial Cable

A continuity test of a short coaxial cable can be performed in the same manner as for a short twisted-wire cable. The ground should be checked from end to end for continuity. A signal lead to ground test should also be performed on every coaxial cable.

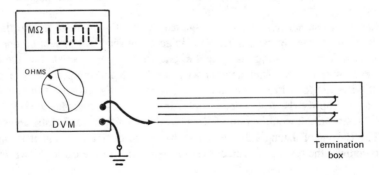

Figure 8-4   Testing of a long cable with a VOM, DVM, or continuity tester.

TABLE 8-2   Wire/Resistance Table

Solid Copper
Ohms/1000 feet
10.15
12.80
16.14
20.36
32.37

Stranded Copper
Ohms/1000 feet
10.32
11.72
13.73
19.80
21.08

Testing of a long coaxial cable presents a special problem in that there is one signal wire and one ground wire. If we short the signal wire to the only other lead (the ground) a continuity test will indicate the same as a shorted cable. For this reason a ground test should be performed first to be sure that there is an "open" between the signal lead and the ground lead.

When the technician is assured that the cable has no internal short between the signal lead and ground, a continuity test can be performed. This is accomplished by shorting the signal to the ground leads (Figure 8-6) and measuring between the two conductors at the other end of the cable. Continuity is indicated for both the ground shield and the center conductor.

These tests are very important for both preassembled cables and those custom made by the technician. A manufactured cable may look perfect but a factory assembly can be faulty. Another version of the continuity checker is a "tone test set".

A tone generator is usually equipped with a BNC, TNC or twisted pair input connector for convenience. Some versions have a mounted light (used for visual check of continuity, load or open circuits) as well as a small speaker. During some open/short test procedures it is helpful to have a test device that can signal an open or short line condition by a beep or tone through a speaker.

An additional device that can greatly assist the wiring specialist is a "shorting cap" (Figure 8-7a). A shorting cap is a male or female cable connector which has the center conductor shorted to the shield or ground. In this manner, the device serves to short the cable at one end. This small device can be plugged into the distant end of a cable or wire group and the specialist can go to the near end and proceed to check for correct cable identity, (with proper equipment) cable length, open/short condition, actual line impedance, and so forth.

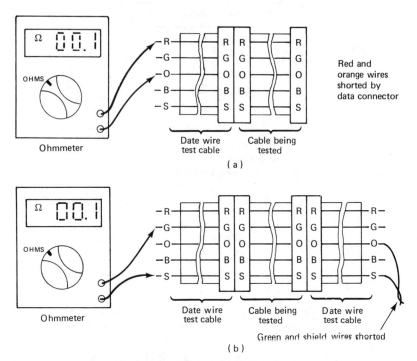

Figure 8-5 Testing a cable for shorts:
a. wire to wire,
b. wire to ground.

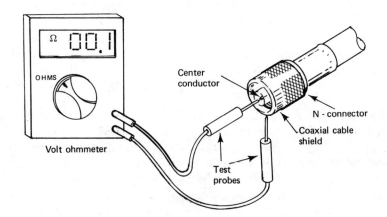

Figure 8-6 A grounded wire test for coaxial cable.

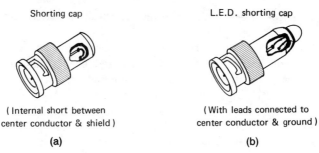

Figure 8-7   Shorting devices:
a. shorting cap
b. LED shorting cap.

A shorting cap can be modified with the addition of a lightemitting-diode (LED) across the center conductor and ground (Figure 8-7b). This device can indicate the presence of low voltage on the line. This feature is very useful in determining "hot" controller ports or pc/workstation connections. It can also check for line continuity when used in conjunction with the tone generator. By placing the modified cap at one end of a cable and the tone-generator at the other, a simple voltage-pass check can be performed. If the LED glows brightly, the line resistance is low and the battery powered generator is operating well. However, if a faint glow appears on the LED, either there is a low-battery condition at the generator or there is some resistance within the line or its connections (this is sometimes called a high-impedance short). No light at all indicates that there is an open circuit existing between the two ends of the cable.

With use and familiarity, the specialist will develop a variety of different cable and/or low voltage tests that will be useful.

Another useful tool is the *inductive amplifier.* One of the most popular uses of this device is in conjunction with the tone generator to locate a "lost" cable within a wiring closet or wire bundle. In this application the tone generator is connected to the "known" end of a wire within a wire bundle and turned on. It generates a signal down the wire. At the other end the inductive amplifier is probed into the bundle, one cable at a time, to find the "lost" wire. When the probe tip is pressed into the correct cable it will be identified by amplifying the tone generator's audio output.

Standardized Testing Procedures (STP) and Online-Database

In order to provide uniform test and timely repair operations, a series of standardized procedures need to be agreed upon and developed. This procedure needs to be established in two different forms. First, an *on-line procedures database* should be established and maintained. Second, a series of *standard procedures manuals* needs to be installed and maintained at each work area and/or technical workstation (as appropriate).

The information to create this document, including procedures and techniques, is obtained from the device-use and device-training manuals, product classes, seminars given on problem determination, and within the personnel of each department. Engineers and specialists with hands-on and/or academic experience should be asked to input their knowledge to add to the manuals.

The contents of these manuals may include installation and testing procedures, troubleshooting guidelines, how to install micro-code on a controller, how to download a program from the host system to perform a special test on a workstation, and so forth. Once all procedural testing outlines are put into the reference manual, and a complete set are agreed upon, an acceptable test should be run. To run the acceptance test, a technician is put into a testing bay with simulated (or real) symptoms and problems. He or she then uses the procedures step-by-step to isolate the problem and determine its cause or if not possible, the next course of action.

There should be a 3-level evaluation on the test method and application:

1) ability to read and comprehend the text

2) ability to perform the testing tasks indicated

3) ability to evaluate the test results to accurately determine the problem cause.

These procedures and guidelines should be documented in two ways; First, a series of departmental reference books should cover the major technical activities that the department is charged with. These manuals should outline what needs to be done at each level and station within the department. The organization should maintain a manual at each work area. Second, a "Test and Troubleshoot" database needs to be built and maintained. This database will outline each work activity and test procedure, including installations, corporate standardization layouts, troubleshooting procedures, and support contacts for additional assistance, escalation guidelines, or carry-over. This "online" database should be available on a 24-hour basis to all departmental employees.

The advantage of this system would be that for the specialist, in the field, the on-line system can be a valuable resource to look up test or troubleshooting procedures or to inquiry the system for the next level and contact of escalation.

Whoever is charged with the responsibility of maintaining the database must also ensure that the a set of "reference volumes" exists within the organization. This individual should be responsible for organizing and updating these manuals.

The final "Test and Repair" manual should be placed in all work areas for the wiring media technician's use. If the technician is unsure how to proceed or unclear about the next step in troubleshooting a problem, the procedures manual should provide direction and guidance in these areas.

8.7 Testing a Fiber-Optic Cable

The testing of fiber-optic cable offers a very different challenge to the technician. Continuity, open, ground, and short test with VOM or an electrical signal source cannot be performed because the glass fiber is an insulation and as such will not conduct an electric current. The fiber must be tested with an **optical time domain reflectrometer (ODTR),** which determines the loss in decibels of the light signal and the length of the cable by measuring the time that it takes the light to travel end to end. An ODTR tester is shown in Figure 8-8(a). The ODTR can be used to determine the overall efficiency of the cable system and any fiber-to-fiber connections within the cable. OTDRS, like TDRS for Copper Media, can show the presence of splices, connectors and impedance mismatches.

While optical time domain reflectometers provide a full range of test and restoration functions for fiber cable systems, they are expensive and difficult to use. The cost usually limits the number that are available for servicing on a network to one and the operation to highly skilled technicians. A somewhat less expensive alternative is a fiber optic fault finder such as illustrated in Figure 8-9. The hand held optical fault finder is user friendly and requires little technical skill to interpret its reading on the CRT.

8.8 Custom-Assembled Cables

Communication cable must often be assembled on site by the wiring technician. Long cables may be installed in place to assure the correct length and then terminated. Great care must be taken in the termination of all three types of communication cable: fiber-optic, twisted pair, and coaxial.

Chapters 2 and 3 should be consulted for the assembly of twisted-pair and coaxial terminals. Detailed instructions for the termination of fiber-optic cable was given in Chapter 4.

Regardless of the type of cable or the type of connector placed on the cable, a complete continuity and ground test must be performed after assembly and the connector should be double-checked to assure that each wire or fiber is connected to the correct pin of the connector.

8.9 Signal-to-Noise Ratio Measurements

The signal-to-noise ratio (S/N) is a communication measurement that indicates the ability of a circuit or electronic system to distinguish between noise, unwanted signals such as cross-talk, electromagnetic pickup, and so on and the selected signal. This ratio is measured in decibels (dB) (see Chapter 1) and a high dB reading is desirable since this indicated the receiver's ability to receive a usable signal.

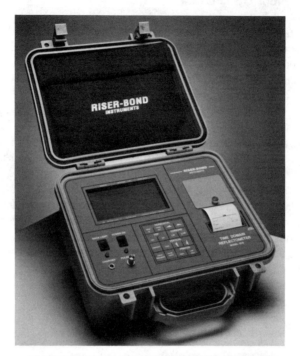

Figure 8-8 A time-domain reflectometer:
a. with an analog or oscilloscope readout
b. with a digital readout.
Courtesy, Riser-Bond Corp.

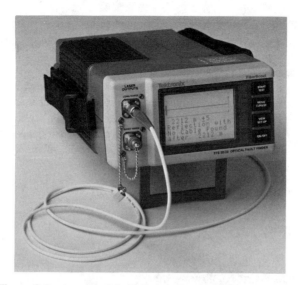

Figure 8-9   An optical fault finder. Courtesy, Tektronix Corp.

A signal-to-noise ratio of 23 to 26 dB is normal on telephone leased lines. This means that the signal to noise is a power ratio of

$$dB = 10 \log \frac{S}{N}$$

where the S/N is the power of the signal/power of noise. For the low 23 dB ratio the power ratio is

$$\frac{23}{10} = \log \frac{S}{N} = 199.5$$

and the high 26 dB ratio is

$$\frac{26}{10} = \log \frac{S}{N} = 398$$

The worst case signal-to-noise ratio of 23 dB represents a power ratio of signal power to noise power of approximately 200, and the high S/N ratio of 26 dB represents a signal power to noise power of 400. We might note that the 3 dB difference of the *S/N* ratio represents double the power. An increase of 3 dB at any level represents a doubling of power, while a power reduction of 3 dB at any level represents a 50% power loss.

A sample of *S/N* to power ratios is given in Table 8-3.

Table 8-4 presents a sample of decibel loss values versus power loss values. The loss of power in a system is denoted as a minus decibel ($-$dB). We refer to all losses and gains in an electrical circuit as *gains* to simplify the calculation of system gain. As an example, suppose that a system had gains of 26 dB and 24 dB and losses of -3 db and -4 db. The overall gain of the system would be calculated by adding the four.

$$\text{system gain} = 24 \text{ dB} + 26 \text{ dB} - 3 \text{ dB} - 4 \text{ dB} = 43 \text{ dB}$$

8.10 Reference Point for Power Level

Early in the development of the field of communication it was recognized that a *universal reference point* to which a system power level of gain or loss could be compared was desirable. Telephone companies chose to use 1 milliwatt, the approximate output level of conversation from a telephone transmitter as the reference. This reference level taken across 600 ohms is called a dBm (decibel 1 milliwatt) and is formulated

$$\text{dBm} = 10 \log \text{signal power in milliwatts}/1 \text{ milliwatt}$$

TABLE 8-3 A sample of S/N ratios as they relate to signal power to noise power.

S/N Ratio (in dB)	Signal/Noise Power
20	100
21	126
22	158
23	199
24	251
25	316
26	398
27	501
28	631
29	794
30	1000

TABLE 8-4 A sample representative of dB losses compared to power losses in electrical circuits.

Decibel Values	Power Ratio (Input/Output)
−1 dB	0.79
−2 dB	0.63
−3 dB	0.50
−4 dB	0.398
−5 dB	0.316
−6 dB	0.251
−7 dB	0.20
−8 dB	0.158
−9 dB	0.126
−10 dB	0.100

The establishment of the dBm allows the reading of a + or − dB reading on a voltmeter scale when a value of 1 milliwatt is inserted into a system.

Telephone companies have established frequencies as well as power levels as standards to which their systems are to be tested. North American companies have established a frequency of 1004 Hz with a power level of 0 dBm (1 milliwatt). Most other telephone companies utilize a standard of 800 Hz at 0 dBm. A reading of 0 dBm at the appropriate frequency in either of these systems means that there is no gain or loss. The 1004 Hz 1 mW reference level is called **transmitted level point (TLP)** in North America and the 800 Hz 1 mW reference level is called *decibel reference* (**dBr**).

A sample test problem is illustrated in Figure 8-10. A generator places a 1004 Hz 1 mW signal on a line which has a 2 dB loss.

8.11 Zero Transmission Level

A system may experience power loss by way of reduced amplification, facility cable connections, or in case of light transmission, a poorly prepared fiber cable connection. To establish the level of reduction in a system as compared to a perfectly functioning system the term **zero transmission level** (dBm 0) is used.

$$\text{zero transmission level} = \text{actual level} - \text{test level}$$
$$\text{dBm } 0 = \text{dBm} - \text{TLP}$$

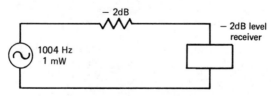

Figure 8-10 An illustration of a TPL test.

The dBm 0 level shows the actual loss in a system as compared to the original specification level. For example, suppose a telephone block has a TLP of 12 dBm but the actual measurements 8 dBm.

Then, $$\text{dBm } 0 = \text{dBm} - \text{TLP}$$

$$= 8 - 12$$

$$= -4 \text{ dB}$$

The system is 4 dB below standard specifications. Using the dB power formula the level of power reduction can be determined.

$$\text{dB} = 10 \log \frac{P_o}{Pi_n}$$

$$= -\frac{4}{10} = \log = \frac{P_o}{Pi_n}$$

$$= 0.399$$

The power level has been reduced to approximately 0.4 of the power specifications.

Table 8-5 is an example of TLP, dBm, and dBm0 measurement in a system.

TABLE 8-5   Measuring the TPL of a System

dB/dBm	Output/input ratio
−4	: 1
−3	: 1
−2	: 1
−1	. : 1
0	1 : 1
1	1.2 : 1
2	1.6 : 1
3	2 : 1
4	2.5 : 1
5	3.2 : 1
6	4 : 1
7	5 : 1
8	6.4 : 1
9	8 : 1
10	10 : 1
16	40 : 1
20	100 : 1
30	1000 : 1
40	10,000 : 1

8.12 The Technical Support Center

A technical support center can help the user through a checklist to determine if the problem exists with the user, the user's application, the data terminal equipment, the data communication equipment, or the link between them and the host. Within this process the software is checked to see if proper communications are established between the terminal and controller to host.

Once the problem has been isolated—if the tech support center cannot fix it—it will be referred on to a higher level of support. When the problem escalation procedure is initiated the analyst at the tech support center will continue to update and monitor problem status. All technical support personnel involved with the problem should continue to update the problem file. Every entry should be time stamped and initialed by personnel. This serves to give a history of how the problem was brought to a solution. The end user and management are updated on problem status.

The information obtained on a problem will serve as a data bank for similar or related problems. Ultimately the information could serve as a diagnostic utility file on a common service system so that other individuals can research a known set of indicators for a possible solution.

The help desk of the tech support center will schedule work, open work orders, track any outstanding problems, review change activity logs for possible service interruptions and/or user service level impact, and issue timely reports for IS and network management. The help desk is typically an administrative and managerial type of job. However, technical requirements can be incorporated into the job description to allow more efficient and competent service at the user's first contact point.

Summary

Although we have classified the protocol level of wiring installation at 0, that does not mean that it is the least important. In fact, the wiring is the base on which all other protocol levels rest.

Questions

1. What are the simplest instruments that can be used for continuity testing of a cable?
2. What is the purpose of a ground test to a cable?
3. What is the difference between the testing of a coaxial cable and a twisted-pair cable?
4. What special problems are there in testing of a fiber-optic cable?
5. How is a signal-to-noise ratio measurement performed on a copper cable?
6. Describe your ideal technical support center.

CHAPTER 9

Documentation of the
Wiring System

9.1 Introduction

Documentation of the cabling system is paramount for effective service to the user community and maintenance of the telecommunication system. Documentation should consist of all the information and data that would be necessary for you to assume the job of telecommunication manager at a new company from which all the technical staff had left.

In other words, documentation must be current, reflecting what is reality at this point in time, and show a history of past events. Current data are necessary for the immediate service to the user community and projection of cable availability. Historical data are necessary to assist in management projection for future data processing needs and as a reference for maintenance and troubleshooting by the technical staff.

Documentation could be almost solely in the form of a database system. However, there should be backup by physical diagrams, charts, blueprints, and forms. Most important to any documentation system is the labeling of all the conduit, feeder cables, terminal blocks, circuits, between building runs, wiring panels access panels, and so on.

9.2 Labeling the Cabling System

Cabling **standards and procedures** should be documented and adhered to for all wiring applications. Cable labels should include such items as

    Building number
    Room identifier
    Room cable number
    Media type

Standards help to establish procedures and present connectivity in a logical and understandable manner. This assists the user community, the technical staff, and management alike by presenting a clear "identifier" for each application. Figure 9-1 illustrates an example of a cable labeling standard.

An example of the foregoing methodology is

    Cable label: 011-1-A26-C93

```
XXX  X - XXX  XXX  XXX
1    1   1    1    1
1    1   1    1    1 ----------------------MEDIA TYPE
1    1   1    1                            B = Broadband
1    1   1    1                            Bk = Backbone
1    1   1    1                            Cxx = Coax
1    1   1    1                              93 = 93 ohms
1    1   1    1                              75 = 75 ohms
1    1   1    1 ----ROOM CABLE              63 = 63 ohms
1    1   1
1    1   1    ex. 000-999
1    1   1        A01-299      C = CAD/CAM
1    1   1                     E = Ethernet cabling system
1    1   1                     Fm = Fiber, multimode
1    1   1                     Fs = Fiber, single mode
1    1   1 -----ROOM           I = IBM cabling system
1    1                         Iw = Inside wiring
1    1                         L = Local area network
1    1                         M = Multiplexer line
1    1                         TN = Teleco cable
1    1                            N = 4 = 4 wire
1    1                            N = 6 = 6 wire
1    1                            N = 8 = 8 wire
1    1                         Ow = Outside wire
1    1                         P = Public address system
1    1                         Tw = Twin axial
1    1 ---FLOOR                Tv = Campus/building TV
1
1
1
1-BUILDING NUMBER

       ex. 000-99
       A01-299
```

Figure 9-1   An example of a cable labeling standard.

In the foregoing example, the cable is located in Building 011, first floor office, room A26, room coax #2. It has an impedance rating of 93 ohms and is a coax-type cable.

9.3 Blueprints and Diagrams

Blueprints are drawings of the physical plant showing the location of all offices, labs, equipment closets, underground tunnels, and so on. Electrical blueprints are drawings of the same plans, but usually have less detail. However, electrical prints may have information not located on building blueprints, such as conduit locations and cable tray locations. Both blueprints and electrical drawings must be kept up to date as spaces are added and deleted from the facility and wiring changes are made to accommodate the user community.

Furniture drawings are made from blueprints but are less useful as individuals have the habit of rearranging their habitat.

An accurate up-to-date set of blueprints showing all main distribution frames (MDF), immediate distribution frames, conduit numbers, cables numbers, and equipment locations should be kept up to date by the technical staff. This documentation shows a physical view of the facility and may be many drawings in a large campus. However, they should be the responsibility of one person and reflect the information in the distribution logs and the database.

9.4 Distribution Logs

The distribution log should be the most up-to-date record of the state of wiring and cabling. This log should reflect the current status of the cabling system and should be filled immediately after any changes are made to the system. In a large campus facility, the distribution log would probably be several files, one for each building or area. An area or building distribution log might be quite simple as shown in Figure 9-2. Here the 20 cables (probably twisted pair) are numbered 1 through 12 by their distribution number. Each is identified as to its origin IDF, cable number, port number, and name of the user. The user's name may be a person, facility, or device (such as a printer).

The simple distribution log may be further detailed by a **wiring closet/controller room log** such as depicted in Figure 9-3. This is one of several worksheets recommended by IBM for the IBM cabling system. Figure 9-4 depicts a further detailed form that is useful to the technician installing or testing the cabling. This drawing depicts the wiring closet and a rack within the closet. The actual location of each terminal block is shown on the drawing.

The drawing documentation in Figure 9-3 is further detailed in the document shown in Figure 9-4. Each attachment to the terminal block is documentation by the type of connection, the accessories in the cable run to the work area, the cable length, and the panel to which the cable is run.

Dist. Circuit No	IDF	Cable No	Ext./Port	Name
1	A			
2	A			
3	A			
4	A			
5	A			
6	A			
7	A			
8	A			
9	A			
10	A			
11	B			
12	B			
13	B			
14	B			
15	B			
16	B			
17	B			
18	B			
19	B			
20	B			

Figure 9-2 A simple distribution log of a small building or a work area within a building.

9.5 Work Area Inventory Sheets

The documentation sheet depicted in Figure 9-2 is used for a work area or work function, such as an office, lab, or a service (such as a research lab)—in other words, a rather small group that has commonality. The form identifies the name of each user within the group, their telephone number, the building, floor, room, cable utilized, distribution number, IDF number, and the types of data processor equipment utilized by the individual. The information for one individual could represent several employees. In a small company the name could represent a function (such as payroll).

9.6 Handwritten Entry Versus Terminal-Based Entry

Historically, organizations have used handwritten entries to update their tracking and inventory systems. Although the setup times was very labor-hour intensive, once in place it worked well provided the facility or inventory didn't often change. However, once change activities began increasing in frequency, the handwritten entry revealed several defects.

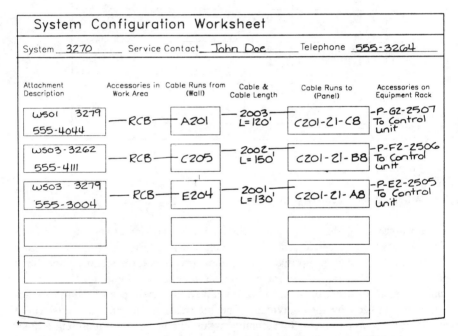

Figure 9-3 Wiring closet/controller room worksheet. Courtesy, IBM Inc.

The most common among them is the problem of maintaining current and accurate records during times of high activity. With moves, adds, and change occurrences, this becomes a large task. Most organizations maintain a master log with copies of it distributed to the various work or activity centers. Usually tracking updates are placed on a department calendar for weekly or monthly reviews, and audit procedures are performed weekly or monthly.

This system works well when things are rather static and running smoothly; however, when the work load is high, the installers or technicians may not have

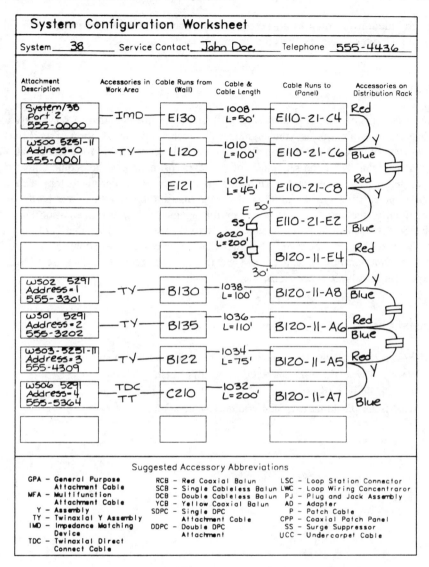

Figure 9-4 System configuration worksheet. Courtesy, IBM Inc.

the time to make log entries. The notes of the changes may become a wad of paper in someone's pocket. Once this begins to happen, management will usually assign the update task to one person. This means increasing head count and adding one more step to the process and, therefore, one more potential failure point.

When the system begins to break down, other tracking and updating alternatives will be proposed and among them will be the cable database management system. This system is a far superior process to update and maintain files as well as provide current data to all personnel who have access authorization. The following chapter discusses the components of such a database.

Summary

The examples of the ideas discussed in this chapter are intended to serve only as ideas for an evaluation of your organization's current documentation or as a starting place for new documentation. The amount of detail necessary for documentation could be debated for the remainder of the text. However, each telecommunication manager must decide what is right for his company considering factors such as age of the existing system, number of cables, distribution systems, other hardware, and finally, the requirements of upper management. Any documentation should also be designed to meet the needs of the wiring system database that is discussed in the next chapter. There must be sufficient documentation to allow the technical staff to function in a timely manner and for the management staff to use to make expedient economical decisions.

Questions

1. What necessitates the establishment of a wiring documentation system?
2. When is a written file documentation system justified?
3. Who should be responsible for a file-based documentation system?
4. What factors could make a file-based documentation break down?

CHAPTER 10

Telecommunication Database

10.1 Introduction

The purpose of this chapter is to outline the requirements of a telecommunication-specific database and to give an example of one such system.

Every wiring or cabling system for a telecommunication network requires some form of tracking facility. The tracking, to be effective, must be capable of update functions (for adding, modifying, and/or deleting records) and retrieval functions (based on request inquiries). The system must also be able to generate inventory and report responses. The "tracking tool" must be capable of allowing the **management information systems (MIS)** departments and telecommunication technical support personal access to the information. Each group or function will have different reasons to access information. Therefore, there should be different access levels controlled by a central person.

- The technical support group will need to be able to do a search based only on the data fields. The only information they may have might be the room information, or the user's name. The system will have to provide the rest of the information such as cable and host connection data.
- Management concerns may include the total connectivity throughout the loading and host channel utilization.
- The control center will have to be able to answer inquiries from users, management, and MIS organizations based upon very little data concerning host service connections, type of cable, and possible overall lengths of specific wire runs.

There are basically two types of *tracking tools* needed to perform the afore-mentioned functions: file-based systems and database systems. These two systems—*file based* and *data based*—will be discussed in the following paragraphs.

10.2 File-Based Tracking System

Traditionally, file-based systems have been used for tracking because of their simplicity and the fact that LANs have been rather small. When the LAN is in a small building or facility the various systems and system connections are simple and a file-based system will probably perform effectively. The decision to select file base or database tracking depends upon the size and complexity of the wiring network (existing and future requirements) and the degree of control that management wishes to have over the system. It also depends on the future product offering or host and/or system requirements.

In the case of a small user community that is fairly stable the file-based tracking method could satisfy both user and the information system department.

As elaborate as file-based system can become, they all share certain inadequacies for a large complex facility with many users. Since there is no common data file, there will be many "copies" of common files. This redundancy carries with it the obvious concern of system-wide accuracy of any one file. Any revision, however minor, means that all copies must be updated. A second concern is in regard to commonality as it is important that all changes carry identical definitions and common factors. Otherwise, a correct update to one file may not be accurately reflected when another file is revised. Also, because many application programs will be accessing various files, there will be some programs with logic overlap. Even with the use of sort/merge programs, time and resources will be expended in different areas to accomplish the same end task.

If the current system cannot keep pace with the growing number of changes and updates or the host service system is being updated on a regular basis to handle the increasing user load then the file-based system may no longer be adequate to meet the requirements.

The information in a company's computer system databases are necessary for employees to perform their work. However, most employees have need for access to only limited amounts of the information contained in a company's data banks. Outsiders should have no need for that information. Some of the information is priority to protect: employees, customer's accounts, and company product development. Therefore, every company should have some form of security system to control access.

Some person or department within the company should be assigned the responsibility of computer and data security. This person should develop the security plans and access control requirements. It may not become this person's responsibility to determine who will have access and to what level. Therefore an entire security access rationale will need to be documented, approved, and implemented. The basic access barrier is a password or passwords assigned to users to allow them to open part or all the system and data banks. Passwords, when properly assigned and used, are a good deterrent to breeches of security. However, they have some drawbacks. For example, users select words that are easy for them to remember such as their wives' names or the make of their car, forgetful users may write passwords down where they can be found and someone can look over a user's shoulder to the monitor when the password is entered, and finally, users leave their system unattended.

For the reasons stated, management should assign passwords and change them often to stress the need for security. Security may also require the shielding of copper cables, locking wiring closets and access doors to prevent the theft of data signals and entry into the data bank.

10.3 Data-Based Tracking System

The **database management system (DBMS)** provides for integrated accessing, linking (implicit and explicit), and managing data files. The database management system provides for a user-friendly interface for nonprogrammers to do inquiries and retrievals easily from shared or common files.

There are three general types of DBMS organizations: hierarchical, network, and relational. For our purposes, the most versatile type will probably be the relational database model. In this model all data are located and written in tables. These "tables" improve flexibility by organizing data in small related groups. When a query comes into the database the table is looked up and data are presented. In the case of a large or complex request, small tables are organized into one large table to meet the needs of the request.

The reason for the small table size format is that larger tables would necessarily have to contain data that would be duplicated elsewhere in other large tables. This duplication of data would add to the overload of the process and would not be an efficient data storage structure.

The DBMS is very flexible and can handle request that it may not have been designed to address due in part to the use of small data tables and implicit and explicit links. The management system will also provide a certain amount of isolation between the data files themselves and the user community through the inquiry language.

10.4 Structured Query Language

In 1986, ANSI defined and standardized the language requirements for relational databases by specifying that **structured query language (SQL)** would be the standard language for all relational database management systems.

SQL is a query language that can be used to create and maintain the database structure. It will also allow data manipulation. SQL allows the user community to specify the data field or group of data requested, and the management system will take the inquiry and search for the requested data from the data tables. Once enough data has been assembled to meet the request, the links are provided and the data provided. Depending upon how the database is structured and oriented, the requested data (which may be only one data field) can be obtained from one or more inquiry panels or menus.

The data tool gives the telecommunications personnel the ability to log onto an online system and perform a number of search inquiries.

Probably two of the most significant uses for such a database are current connectivity status and forecasting. Connectivity status informs the telecommunication personnel as to what is connected to the system, how it is connected, and to whom it is assigned. The forecasting capability allows the communications manager to know exactly the number of cables in the system, their current usage, and their future capacity.

10.5 Basic Components of a Database Management System

There are three basic components to the database management requirement for a telecommunication system. These are

1. resources required to plan, write, and maintain the database
2. a database administrator (DBA)
3. a contact point for all telecommunication requests for cable or wire connection to host services

Although the initial cost of creating a database management system is greater than that of a file system, its strengths far outweigh the additional expense considerations. Some of the advantages of the DBM system over the file system are

1. The database offers greater access to organized data than the file system through the use of host and query languages.
2. Data can be accessed from either applications program or query languages that can be written for the nonprogrammer. This ease of use helps all concerned to obtain the data or records needed without the requirement of advanced language skills.
3. The user can log onto an interface that will make inquires on the database for the retrieval or update information simple.
4. The database can be used for technical support as well as planning and management for the proper and timely coordination of premise wiring systems.
5. Not only will the database provide an accurate picture of current usage, but it will also provide information on the current product mix of PCs, workstations, printers, and terminal use.

10.6 Database Manager's Responsibility

The database manager is responsible for the database. He or she must ensure the accuracy of the data for tracking, inventory, and forecasting purposes. The DBM may have been involved with the writing of the program and/or the organization of the files or records that built the system. In either case, the DBM must ensure the effectiveness and efficiency of the data retrieval system.

The DBM has the authority to assign read-only or read-and-write access to the database as management and technical support require different levels of access to the database depending upon their job assignments.

The DBM is responsible for setting minimum performance measurements including system response time.

The DBM may have been or may become involved in the writing of the database program or program requirements. Regardless, the DBM must assure that

1. the system be designed with enough flexibility to evolve and accommodate upgrading with a minimal impact on service levels
2. a contact point be maintained by the technical support center in order to account fully for all requests for new service and/or cable runs
3. the database be used by both technical support as well as planning and management for the proper and timely coordination of premise wiring systems
4. the database provide an accurate picture of current product mix of PCs, workstations, and terminals in use

10.7 Sample Database

To show how a sample database can be constructed and what it will look like to the user community, technical support center, and management, some basic assumptions about the design and layout are made for the purpose of explanation.

This example is a sample only. In the real world there would be several meetings between affected and interested groups to determine the database design, how it would be laid out, who would have access, and myriad other "individual" and "corporate" concerns, special requirements, and needs.

The first is to determine if the system will be user-data driven or host-service driven.

A user-data driven facility would accommodate mixed hardware such as Apple, IBM, HP, and/or DEC workstations all connected to the network. In other words, the user controls the device hardware, and the host provides a generic *device-tolerant service platform.*

In a host-service driven facility, the hardware and software are predetermined for the community. Any workstation or terminal hookup is specified to a certain group of hardware. All users coming into the facility would use IBM, DEC, HP, or Apple workstations. The host dictates the protocol. There would be no opportunity to connect mixed vendor PCs or workstations on the existing network.

Our database example will be menu or screen driven. It will consist of five major panels to be used for inquiry, data retrieval, updating, and initiating a work request.

- First screen will be the master inquiry panel.
- Second screen will be the location/room cable panel.
- Third screen will be the link and connect panel.
- Fourth screen will be the service availability panel.
- Fifth screen will be the work order panel.

The master inquiry panel (Figure 10.1) is a read-only file, although all fields are searchable within the panel.

The original data for this panel should have been obtained from verified records or by a physical inventory of the existing facility and an accurate documentation of all cable connections from the host to the end user's room termination by the department in charge. This allows various organizations to have access to this panel for information and status update or related activities within their functions. Users within selected organizations can directly access this panel and have restricted write access so that they can submit requests online. The panel can also be used further for the following purposes:

- By support center personnel to schedule work, track a request for service or a device, and initiate host or remote system connections.
- To monitor current connectivity loads by system. This will help in anticipating bottlenecks and possible performance problems.
- To support center personnel to schedule work, track a request for service or device, or initiate host or remote system connection request.
- For management to inquire about the availability of existing cable runs within a given department or number of offices. This may be for resource consolidation (double up personnel in offices) or upgrade service levels in a particular group.
- For management approval when an employee needs access to a secure system or increased service support levels that need prior management approval.
- For special requirements. For example, suppose there is a litigation department that is dealing with leased lines for specific Japanese products. These product offerings will be designated for the overseas market, and all applications programs will be written for Japanese-speaking clients.

The development computers will need to run under Japanese language requirements. Because of this, a small number of controllers will be specifically configured and adapted for that purpose. However, to set up an application system both the controller and the workstation have to be specified or the program will not run properly. In this example the requester would specify which device is needed, which control must be connected, and so on.

10.8 Field Description of Master Panel

The panel is broken into eight categories:

User data
Room cable data
Terminal/device data
Type of request
Work activity section
Type of connection required
Type of host required
IS only

- Field 1: User's name
This field is reserved for the user's name or contract. It should be entered as the last, then first name. In this format, a search routing or inquiry can easily locate the correct user. This field includes the user's telephone number. If a work request is called in for a new hire or contractor, the name field may not be known. However, the office and phone number should be available in case additional information is needed. Also included is the company I.D. number.

- Field 2: Location
This field consist of both the building number and room number. In a search routine based solely on room number, the analyst should be able to obtain the following information:
1. Number and type of cables existing in room presently.
2. Type of host and controller service available on which room cables.
3. Type of terminals and/or devices already in place in the room. Typically, this is the most often used search to provide information to the requester.

- Field 3: Department number
This field is really for tracking purposes. For example, if a department were physically located within one building, or perhaps, on one floor or core area, all host/controller connections could be located and routed through common controller services. A good example of this would be a litigation department or a project design team. All online resources could be located on a single (or few) dedicated controller(s). This dedication has some immediate advantages of security and commonality to the functional unit.

MASTER INQUIRY PANEL

USER DATA

1) USER NAME: (LAST) _____ (FIRST) _____
   USER'S OFFICE PH: _____ COMPANY I.D. _____
2) LOCATION: (BLDG) _____
   ROOM NUMBER: _____
3) DEPT NUMBER: _____ DEPT MANAGER: _____
   DEPT MANAGER PH: _____

ROOM CABLE DATA

4) NUMBER OF CABLES IN ROOM:
   ( ) 1,   ( ) 2,   ( ) 3,   ( ) 4,   OTHER ( ): CABLE # _____
   NUMBER OF CABLES UNUSED IN ROOM:
   ( ) 1,   ( ) 2,   ( ) 3,   ( ) 4,   OTHER ( )
5) TYPE OF ROOM CABLE AVAILABLE:
   CABLE #1: TWISTED PR ( ),   COAX ( ),   FIBER ( ),   LAN ( )
   CABLE #2: TWISTED PR ( ),   COAX ( ),   FIBER ( ),   LAN ( )
   CABLE #3: TWISTED PR ( ),   COAX ( ),   FIBER ( ),   LAN ( )
   CABLE #4: TWISTED PR ( ),   COAX ( ),   FIBER ( ),   LAN ( )
6) ROOM CABLE #1 IDENTIFIER: _____
   ROOM CABLE #2 IDENTIFIER: _____
   ROOM CABLE #3 IDENTIFIER: _____
   ROOM CABLE #4 IDENTIFIER: _____

TERMINAL/DEVICE DATA

7) NUMBER OF TERMINALS AND/OR DEVICES IN ROOM:
8) TERMINAL/DEVICE TYPE, SERIAL NUMBER (S/N) AND MODEL
   DEVICE #1 TYPE: _____ S/N: _____ MDL: _____
   DEVICE #2 TYPE: _____ S/N: _____ MDL: _____
   DEVICE #3 TYPE: _____ S/N: _____ MDL: _____
   DEVICE #4 TYPE: _____ S/N: _____ MDL: _____

TYPE OF REQUEST SECTION

9) NEW SERVICE: ( )
10) RECONNECT SERVICE: ( )
11) DELETE EXISTING SERVICE: ( )

Figure 10-1 Master inquiry panel.

WORK ACTIVITY SECTION

12) INSTALL: ( )
13) EQUIPMENT NEEDED: _____

_____

14) ESTIMATED TIME OF ARRIVAL (ETA): _____
15) REMOVAL: ( )

TYPE OF CONNECTION REQUESTED

16) DIRECT CONNECT (HARDWIRED): ( )
17) TERMINAL CONCENTRATOR: ( )   IF CHECKED CONC. S/N: ____
18) FRONT END PROCESSOR: ( )   IF CHECKED FEP S/N: _____

TYPE OF HOST REQUIRED

19) FLOOR SYSTEM (NATIVE): ( ) _____
20) REMOTE SYSTEM: ( ) _____
21) DEDICATED SYSTEM: ( ) _____
22) ISOLATED SYSTEM: ( ) _____
23) ENGINEERING SYSTEM: ( ) _____
24) TEST SYSTEM: ( ) _____

I.S. INTERNAL USE

25) RESTRICTED ACCESS: YES ( )   NO ( )
26) AUTHORIZING MANAGER: _____   PH: _____
27) CONTACT NAME: _____
28) CONTACT PHONE NUMBER: _____

END OF MASTER MENU

Figure 10.1   (*Continued*).

- Field 4: Room cable data
  This information, once input into the system, will allow for specific searches to be done on a particular cable, coax, or fiber line.
  On the query screen, the rest of the data will be presented. In this manner, questions concerning whether or not a particular cable is active, what service is available (if any), and so on can be addressed quickly.
- Field 5: Room cable type
  Specifies the type of room cable available.

- Field 6: Cable labeling information
  Provides corporate cable labeling information.
- Field 7: Terminal or devices
  This field shows the number of devices in room.
- Field 8: Terminal or device data
  This field will provide the numeric identifier for the particular device or terminal in the office from which the request originated. Information given is:
  Serial number—This is the serial number of the device or terminal. Fields 10 and 11 are excellent resources to conduct a quick audit of physical assets within a department or function. Accounting departments may want to get a monthly or quarterly report with at least these two fields included.
  Model—This will be the model of the terminal or device in the user's office or laboratory.

10.9 Type of Request Section

- Field 9: New service
  This selection is used for new service. When marked with an "X," this begins the new service work activity. The analyst and tech will work off this order as well as the room cable panel and connect panel to complete the installation service.
- Field 10: Reconnect service
  This field is used for reconnecting a deleted service.
- Field 11: Delete existing service
  This field is used to delete a service when, say, an employee is vacating an office or lab.
- Field 12: Install
  This is an installation order. Once this box is checked all telecommunication personnel will know that this is an order for hardware installation.
- Field 13: Equipment needed
  This is for hardware needed such as terminals, printers, and so on.
- Field 14: Estimated time of arrival (ETA)
  This is a data field. It is used when equipment or devices are being sent in from different locations. It can be as specific as the actual arrival data at the user's office or room or it may be only a guess as to when the needed items will arrive.
- Field 15: Removal
  If checked with an "X" this request is to remove a piece of equipment or device. The analyst and tech support group will have to set up some form of tracking, inventorying, and storage facilities for this purpose.
- Field 16: Direct/connect hardwire
  This field is used to identify those devices that require direct connection or hardwiring, usually to a floor or native system.

- Field 17: Terminal concentrator
  This field is used to identify a request for a terminal concentrator connection.
- Field 18: Front end process
  This is the main processor.
- Field 19: Floor native
  When marked, this indicates the user requires a floor system connection.
- Field 20: Remote system
  When marked, this indicates the user requires a remote system connection.
- Field 21: Dedicated system
  When marked, this indicates the user requires a dedicated system connection.
- Field 22: Isolated system
  When marked, this indicates that the user requires a connection to a stand-alone system.
- Field 23: Engineering system
  When marked, this indicates that the user requires a connection to an engineering system.
- Field 24: Test system
  When marked, this indicates the user requires a connection to a test system.
- Field 25: Restricted access
  There can be a question of security regarding access for telecommunication personnel to the user's office or room. If there is a case of restricted access, this field should be set to yes and field 8 (contact) should be filled in with that person's name as well as field 9 (contact phone number) for the phone number of the person to contact.
- Field 27: Contact
  This is the person to be contacted for access to the requester's room or office. This field is used when the restricted access concern or the user is out of the office and this person has a key to an access approval.
- Field 28: Contact phone number
  This is the phone number of the contact.

Specific controller and/or port requirements are other areas of unique information to specify a controller that is configured on a certain microcode level. The port information may be requested for special PCs or workstations that need to be port configurated or customized to meet a unique device or terminal requirement.

For facilities that have three or more point network strategies, the following panels will assist in completing the connection paths and links between user device and host.

To complete this example, we must take some liberties with the fictitious facility that is being described. For instance, there is a minimum of one and a maximum of five cables in each office or laboratory space. The location/room cable panel (Figure 10-2) can be used quite effectively when the facility is primarily host-service based. The services that are available are indicated within each office or lab space. If enough system hardware is available, a "hot service" concept can be implemented to provide at least one live service connection within each room on a designated room cable for service connection to user's device.

The major tracking point within this strategy is the room cable. The room cable data is important because it allows any person with read-only access to check on service level availability by room cable data. Some organizations will only track "connect points" or host-to-room cable links only and disregard device hardware altogether. In this instance the primary concern is with controller connections, controller microcode levels and device support grouping within a defined product range. An appropriate database and support service is developed to maintain the system.

The location/room cable panel (Figure 10-2) is the second screen panel in the database series. The panel contains the following:

- Field 1: Building number
  This format is the same as for the master inquiry panel.
- Field 2: Floor
  This field is a three-digit field for alphanumeric data.
- Field 3: Room number
  This format is the same as for the master inquiry panel.
- Field 4: Room cable number
  This field is a four-digit alphanumeric code. There is a total of five fields to support up to five cable runs within the same office or lab space.

LOCATION/ROOM CABLE PANEL

1) BLDG NUMBER: _____

2) FLOOR: _____

3) ROOM NUMBER: _____

4) ROOM CABLE NUMBER: _____

5) CABLE TYPE: _____

6) CONNECTION COUNT: _____

7) CONTROL UNIT TYPE: _____

8) CONTROL UNITE SERIAL NUMBER: _____

9) HOST SYSTEMS: _____

10) WORK REQUEST ORDER NUMBER: _____

Figure 10-2   Field description for location/room cable panel.

Example: cable 110 = −110

cable 02 = −002

cable 14/7 = −14/7

With the information provided by the building, room and cable number fields, the technical support personnel can go directly to the cable selection panel for all room cables run into this room and determine which ones are active.

- Field 5: Cable type
Reference "Cable labeling Chapter 9."
- Field 6: Connection count
This field is optional. When filled in it will give the number of connections between the controller and the end user's office. Another variation of this is the connection count between host and end user's office.
- Field 7: Control unit type
This information will assist management and technical support in tracking connectivity base. It also gives information concerning the type of controller and microcode level to determine support service level for device or product, LAN or WAN, and so on.
- Field 8: Control unit serial number (if applicable)
The serial number of the controller can be input to specify a specific controller.
- Field 9: Host system
This can be information concerning a terminal concentrator or a direct connect, dedicated system.
- Field 10: Work request order number
If a work order is indicated then a request may be outstanding for a hookup or service connection.

The room cable panel is the list of all cable numbers assigned but not always present in the room. It is important to note that cable room numbers can start at any point since we have designated that all room cables carry a standardized cable number and type of cable (Chapter 9).

10.10 Field Description for Link and Connect Panel

The third screen (Figure 10-3) in the series contains the following:

- Field 1: Building number
This format is the same as for the master inquiry panel.
- Field 2: Room number
This format is the same as for the master inquiry panel.

LINK AND CONNECT PANEL

1) BLDG NUMBER: _____

2) ROOM NUMBER: _____

3) ROOM CABLE NUMBER: _____

4) HOST SYSTEM: _____

5) CHANNEL FIELD: _____

6) CONTROL UNIT TYPE: _____

7) CONTROLLER SERIAL NUMBER: _____
   MICROCODE LEVEL: _____

8) CONTROLLER FLOOR LOCATION: _____

9) CONTROLLER PORT ADDRESS: _____

10) PORT CONFIGURATION: _____

11) CONTROLLER BREAKOUT PANEL RACK NUMBER: _____

12) RACK AND FIELD POSITION: _____

13) PANEL TERMINATION: _____

14) COMPUTER FLOOR TERMINATION: _____

15) REMOTE ROOM PANEL: _____

16) WORK ORDER NUMBER: _____

Figure 10-3   Link and connect panel.

- Field 3: Room cable number
  This format is the same as for the master inquiry panel.
- Field 4: Host system
  This will be the actual dedicated host system or terminal concentrator.
- Field 5: Channel field
  This field is optional. When filled in it indicates the number of the channel to which the controller is attached.
- Field 6: Control unit type
  There are various controller types each with specific characteristics.
- Field 7: Controller serial number
  This is the controller's serial number
- Field 8: Controller floor location
  This is a six-character field for alphanumeric data describing the physical location of the controller.
- Field 9: Port address
  This is the actual port address as defined by the system.
- Field 10: Port configuration
  This is the port definition for this specific port.

- Field 11: Controller breakout panel rack number
  This is the location of the panel where a 32-port controller is broken out into 32 individual port terminations.
- Field 12: Rack/field position
  This is the location of the above-referenced panel, if the panels were put into a subrack system.
- Field 13: Panel termination
  This is the actual port terminal on the rack or field panel.
- Field 14: Computer floor terminal
  This is the location of the computer cable terminations. In a rack- or field-mounted system, this is the host panel.
- Field 15: Remote room panel
  This is the location of the panel where the remote room cables terminate. In this case the remote panels are not located at or in the computer and controller floor area but somewhere else in the facility.
- Field 16: Work order
  If filled in, this is the work order associated with this cable run.

10.11 Service Availability Panel

The service availability panel (Figure 10-4) is used for data retrieval. The cable data will be entered in the request portion of the panel. This information will be known from the location/room cable panel.

In response to the request, the database system will provide the reply data. This information will include the controller type, its serial number, the controller microcode level, and the system host that is on the specified cable number. The other portions of this panel are repetitive for those rooms or labs with more than one cable. If there is not a host-service connection on a particular cable that will also be noted.

The service availability panel will display up to three cable-to-host connections in our example. The program could be written for as many cables as required. This panel is very useful to both support personnel and the user community because each can look up current status of host services by the cable number.

10.12 Field Descriptions for the Work Order Panel

The fifth panel is the work order panel (Figure 10-5). This panel is based on the master inquiry panel, except that this is a read/write panel. Although it has much of the same information duplicated on it from the master inquiry panel, it differs in that the analyst can, through the handshaking routine of other data tables,

```
SERVICE AVAILABILITY PANEL

   REQUEST

#1) CABLE NUMBER: _____
                              REPLY TO REQUEST #1
                              CTLR: _____   MODEL: _____
                              S/N: _____
                              MICROCODE LEVEL: _____
                              LOCATE COORDINATES: _____
                              TOTAL # OF CNTLR PORTS: _____
                              TOTAL # OF AVAIL PORTS: _____
#2) CABLE NUMBER: _____
                              REPLY TO REQUEST #2
                              CTLR: _____   MODEL: _____
                              S/N: _____
                              MICROCODE LEVEL: _____
                              LOCATE COORDINATES: _____
                              TOTAL # OF CNTLR PORTS: _____
                              TOTAL # OF AVAIL PORTS: _____
#3) CABLE NUMBER: _____
                              REPLY TO REQUEST #3
                              CTLR: _____   MODEL: _____
                              S/N: _____
                              MICROCODE LEVEL: _____
                              LOCATE COORDINATES: _____
                              TOTAL # OF CNTLR PORTS: _____
                              TOTAL # OF AVAIL PORTS: _____
          END OF SERVICE AVAILABILITY PANEL
```

Figure 10-4   Service availability panel.

assign controllers, specify host systems, and choose from other options on the panel. Finally, once all the required data fields are completed, the tracking database will assign a work order number.

This information panel is filled out by the analyst who runs the support center. Once all the required data fields are filled in, the system will assign a work order number. This number will be used to track the work to be performed with time-stamped entries from the various individuals or groups to which the work is assigned.

WORK ORDER PANEL

ORDER #: _____

USER DATA

1) USER NAME: (LAST) _____   (FIRST) _____
2) LOCATION: (BLDG) _____
3) ROOM NUMBER: _____
4) ROOM CABLE NUMBER: _____
5) DEPARTMENT NUMBER: _____
6) OFFICE PHONE NUMBER: _____
7) DEPT NUMBER: _____   DEPT MANAGER: _____
                                 DEPT MANAGER PH: _____

ROOM CABLE DATA

8) NUMBER OF CABLES IN ROOM:
   ( ) 1, ( ) 2, ( ) 3, ( ) 4,   OTHER ( ): CABLE # _____
   NUMBER OF CABLES UNUSED IN ROOM:
   ( ) 1, ( ) 2, ( ) 3, ( ) 4,   OTHER ( )
9) TYPE OF ROOM CABLE AVAILABLE:
   CABLE #1: TWISTED PR ( ), COAX ( ), FIBER ( ), LAN ( )
   CABLE #2: TWISTED PR ( ), COAX ( ), FIBER ( ), LAN ( )
10) ROOM CABLE #1 IDENTIFIER: _____
    ROOM CABLE #2 IDENTIFIER: _____

TERMINAL/DEVICE DATA

11) NUMBER OF TERMINALS AND/OR DEVICES IN ROOM:
12) TERMINAL/DEVICE TYPE, SERIAL NUMBER (S/N) AND MODEL
    DEVICE #1 TYPE: _____   S/N: _____   MDL: _____
    DEVICE #2 TYPE: _____   S/N: _____   MDL: _____

TYPE OF REQUEST SECTION

9) NEW SERVICE: ( )
10) RECONNECT SERVICE: ( )
11) DELETE EXISTING SERVICE: ( )

Figure 10-5   Work order panel.

12) INSTALL: ( )
13) EQUIPMENT NEEDED: _____

14) ESTIMATED TIME OF ARRIVAL (ETA): _____
15) REMOVAL: ( )

TYPE OF CONNECTION REQUESTED

16) DIRECT CONNECT (HARDWIRED): ( )
17) TERMINAL CONCENTRATOR: ( )   IF CHECKED CONC. S/N: ___
18) FRONT END PROCESSOR: ( )   IF CHECKED FEP S/N: _____

TYPE OF HOST REQUIRED

19) FLOOR SYSTEM (NATIVE): ( ) _____
20) REMOTE SYSTEM: ( ) _____
21) DEDICATED SYSTEM: ( ) _____
22) ISOLATED SYSTEM: ( ) _____
23) ENGINEERING SYSTEM: ( ) _____
24) TEST SYSTEM: ( ) _____

I.S. INTERNAL USE

25) RESTRICTED ACCESS: YES ( )   NO ( )
26) AUTHORIZING MANAGER: _____ PH: _____
27) CONTACT NAME: _____
28) CONTACT PHONE NUMBER: _____

SPECIFIC SYSTEMS, CONTROLLER, PORT OR CONFIG REQUIREMENTS

_____
_____

WORK ORDER DATA

WORK ORDER NUMBER: _____ DATE ASSIGNED: _____
RESPONSIBLE DEPT: _____ ESTIMATED TURNAROUND DATE: _____

Figure 10-5   *(Continued)*.

Summary

The development of an effective complete data-based system for all cabling, data equipment type and location, user location and identification, maintenance, system utilization, and growth projection is a necessity for the efficient operation of a telecommunication system. Such a system can help to "smooth out" the daily problems that happen with any telecommunication system that seem to be trying to drive the users, the technical team, and the management crazy.

However, it is a fact that such a database will allow for better utilization of existing cabling, equipment, and time of company personnel. These factors overshadow the possibility of a harmonious work force as they can be related directly to profit and loss.

Questions

1. What is the purpose of a wiring database?
2. Identify the minimum functions that you would require in a database.
3. Compare the host-driven and user-driven systems.
4. When would a tracking data base not be needed?

CHAPTER 11

Management and the Wiring Management Problems

11.1 Introduction

The telecommunications manager is faced with many problems of changing business dimensions. These include, but are certainly not limited to, the following list:

1. Evaluating existing voice and data networks to assure that they meet current and future needs
2. Office moves, adds, and changes
3. Growing network and hardware (PC, workstation, etc.) requirements
4. Changing network services (expansions and upgrades)
5. Company growth as well as resource consolidations
6. Staying current with new product offerings and evaluating them in terms of cost/benefit
7. Making an economical business decision for new equipment and products or the upgrade of current equipment

8. Presenting the decision in point 7 to upper management

9. Introducing new technologies and products into the connectivity network with the lowest possible integration impact

To compound the problems, the end user's requirements (imagined and real) are a moving target for the telecommunication manager. The current inventory of user products and devices can range from major brand PCs (IBM, HP, DEC, etc.) to PC clones (with varying degrees of compatibility), workstations, specialized printers, multifunction phones, multiplexers, plotters, multiple-host service or special dedicated host services, LANs, Ethernets, dumb terminals, intelligent terminals, Cadam and high-level graphics terminals/PC, to specialized language devices. The list could go on.

Each of these devices has specific and, many times, different wiring and cabling requirements for interconnect and different host-service requirements.

Historically, the premise cabling system has gone unrecognized as a separate resource. Typically the cabling requirements and the associated cost were developed as a part of, or requirement for, another product or device. The separate cabling system was rarely thought of as a product itself.

The wiring and cable system for premise and off-premise interconnect networks is an area that has been overlooked by management. Because of this, management and technical functions lost sight of the reuse value of the existing interconnect network already in place. Corporate America has failed to identify the significance of currently installed wiring systems. Existing cabling networks can make a contribution through many product-development cycles.

This oversight is not surprising when the areas of responsibility are often ill-defined for the telecommunication manager. It is the opinion of the authors that managers have a three-part problem.

First, the business rationale for specific hardware and software products must be determined. Business justifications must be put together and complemented by an overall plan toward greater utilization efficiencies of current resources. In view of the current state of technology, there is more than a good chance that any product purchase will be outdated prior to its life cycle, and therefore an argument could be made for staying out of the current market and waiting. This argument can always be made about any technically based product. However, with good managerial direction based on a clear no-nonsense approach to telecommunications networks, this argument can be nullified.

Second, once all equipment has been ordered and installed, the impact upon the existing system must be understood in terms of system performance versus service interruption. In many data processing shops, any service level deterioration cannot be tolerated except in emergencies. Management must be able to qualify both service interconnection time (and place this time requirement during off-prime shift period) and value-added function or service.

Third, all product installation, including interconnect networks, must be inventoried and "tracked" as part of an ongoing process to fully document the data processing facility. Once the documentation is up to date and current, product and floor planners can take better advantage of all "systems" and "networks." Both components are resources that can be upgraded to meet future requirements, but without proper documentation, and therefore, measurement points, the resources at hand cannot be fully utilized.

11.2 Tracking

Current wire and cable systems need to be documented and tracked. This should be done with proper documentation forms and on a database program (Chapters 6 and 9).

11.3 Measurement, Testing, and Troubleshooting

When the wiring system or any part of a wiring system is changed, it must be tested by the appropriate technical personnel (Chapter 7). Once the interconnect networks are tested and documented, measurement points can be established and monitored to determine system loads, application factors, and areas of underutilization.

11.4 Retrofitting

When retrofitting or upgrading a facility, the concept of "highest possible telecommunication application" should be a bullet item in all meetings and discussion groups. That is, the concept of maximum utilization of current or projected facilities, whether it is a small office space or a large laboratory building, should be the central theme of discussion.

11.5 Cost Factors

Cost factors should be balanced against current facilities' utilization and possible future applications. In this manner the additional cost for more function can be determined in advance versus the estimated cost of "upgrade" retrofit at some future data.

11.6 Database and Database Development

Some thought must also be given to the database and database development. It must be determined if the facility is to be user-data driven or host-service driven.

The answer will have far-reaching effects on the level of service provided and the type of database that is written.

Questions that must be answered so as to determine the proper type and level of database are

- Will the IS and telecommunications departments track only cable runs and the cable host-service levels irrespective of the user community, or will the responsible departments decide to control all data fields, connection points, and devices with the network? What part of the interconnect network will the telecommunications department have control over? What part will the IS department have?
- Who will take the responsibility of performing overall quality checks and standards for system performance interconnect network control, maintenance, and supervision?
- Who will manage the cable networks, including premise and off-premise networks?
- Who will install, update, and maintain LANs and all new telecommunication technology?
- How will it "track" network usage? What level of service is to be provided and in what areas? What provider service and turnaround time is to be allowed?

Obviously any telecommunications tool will be able to track all data fields, but it is far more complex for a single department to control and maintain user devices, user connections, host-service access levels, and wire and cable management. However, the answer to these questions will set in motion the structure of the telecommunications department's responsibilities and the requirements for the entire user community.

11.7 Writing a Bid Proposal Request

It is the responsibility of the TCM to write or have written a concise detailed bid proposal request (Chapter 11). The document must be exact in terms of work to be accomplished, type of materials and equipment to be used (if either or both are important), time requirements, company's responsibilities, contractor's responsibility, and safety responsibility.

The writer must know exactly what it is that is being requested, so that there is no possibility for confusion on the part of the bidder. If money is the primary concern, specifications of a specific manufacturer's products should be avoided. A specific manufacturer's specification should only be used when no other equipment can possibly perform the tasks.

Summary

The position of the telecommunication manager within a company requires that a person have a wide scope of knowledge based on both education and experience. The person in this position should have a strong technical background as well as managerial skills. This is not a position for on-the-job training.

Questions

1. Who should plan the wiring system?
2. Who should be responsible for tracking the wiring?
3. Who should be responsible for evaluation of the cost factors relating to retrofitting?

CHAPTER 12

Writing the Specifications
for a Bid Proposal

12.1 Introduction

Preparing the specifications for an **information for bid (IFB)** task is one that few telecommunication personnel would seek. This task is usually placed on about the same level as that of documentation and layout of the wiring system. Like all documentation, it is very important to the well-being of the communication system. An IFB also has ramifications that are reflected in the morals of the user community and the profit and loss statement.

A poorly written IFB can result in incorrect installation of cabling, installation of the wrong type of cable, and/or installation of the cabling at the wrong place. This can cause system down time, poor productivity, and outright stoppage of a user's station or the complete office.

12.2 Details to Include in a Request for Bid Proposal

The request for bid proposal should be written by someone whose writing is meticulous. The writer must collect information from all the parties involved in the project so that it is fully understood. This should take place well in advance of the proposal release so that there is time to understand fully the needs of all parties, write the proposal, and have the interested parties review the proposal and rewrite the document to include any concerns by these parties.

The RFP writer, to develop a reliable document, should take the following steps in preparing the request:

Research

There must be sufficient lead time for the writer to research the project before writing begins. This research should include the following:

1. Contracts and/or meeting with all interested parties to determine exactly what is needed/wanted and have these needs/wants approved by upper management.
2. Determination of the types of cabling that is presently installed in the affected areas, and the type of cabling that is required for any new equipment that is included on the approved need list.
3. Determination from the planning department or product vendors of the types of cabling and terminations that are necessary for any new equipment.
4. Determination of the exact description of any specific products so that these can be specified to the contractor. The writer must be specific or the vendor may install something that meets the specifications of the proposal but not the needs of the users.

12.2.1 Special Facility Consideration

1. *Equipment Rooms*—Are there equipment rooms available, and are they of sufficient size? It is false economy to have an equipment cabinet stuck in a corner of an office or hall. One can be sure that the needs for connections will grow with time.
2. *Ventilation*—Is there proper ventilation in the equipment rooms and the offices/laboratories to accommodate the new installation?
3. *Lighting and Power*—This is an area that is the most often forgotten. Users seem to think that systems and equipment can be continually added to the ac voltage outlets or that a new line can be run in from the power panel. Accumulate the power requirements for all the new equipment that is to be installed, and present the data to the plant electrical department for a review of the power availability. The cost of any additions should be considered *before* any contract proposal is written. This additional cost may change the "needs" of the user group. The authors have experienced situations where the cost of additional power far exceeded the cost of the improvements to the data communication system. Figure 12-1 depicts an example of a request for bid proposal (RFP).

12.3 Development Time

The writer must allow sufficient time

1. for meetings before the actual writing to collect information from the user community
2. to review the information and perform the actual writing
3. to have the user community review the completed document to assure that what they said they wanted is what they really want (probably by now their needs have changed)

12.3.1 Contractor Response Time

Time must be allotted to assure that all contractors have enough time to evaluate the project and prepare a proper bid

This allotment should allow the writer and contractor time:

1. to get the contract proposal into the hands of the appropriate contractors
2. to allow the contractors to review the proposals fully and respond with bids
3. to review the bids and compare the contractor's responses to the bid proposal

12.3.2 Installation, Testing, and Evaluation Time

A contract would not be signed unless the user community had pressing needs for new products. However, the telecommunication personnel must not let this factor diminish the need for time to allow the contractor to perform the actual work, to test, evaluate, and perform acceptance tests of the installation.

12.4 Special Contractor Considerations

Before submitting a bid proposal, the writer should review the document from the point of view of the contractors. He or she should ask the following questions of the proposal:

1. Is the proposal specific on every item that is requested or is any part vague?
2. Does the proposal specify unit pricing where appropriate? Whenever possible, specify separate pricing for material and labor. This will allow you to make a better comparison of the separate bids. Bids that are too low in either area should be suspect.
3. Does the proposal specify a deadline for bid response? A bid response time should be specified with date, year, time of day and place. For example, at 1200 hours, July 5, 1991 at the office of the president, San Jose City College,

2100 Moorpark Avenue, San Jose, California 95128. Specify that it is the responsibility of the contractor for mailed bids to meet the time deadline.

12.5 Bid Proposal Forms

You may wish to develop specific forms for both request for proposals and bid response.

Figures 12-1 through 12-9 are examples of detailed wiring documentation forms utilized by IBM in the *Using the IBM Cabling System with Communication Products.* This and other IBM telecommunication publications are available from the address given in the vendor section of the Appendix.

A sample of a step-by-step detailed work proposal is presented in the following paragraphs.

12.6 Detailed Work Proposal

The detailed work proposal (DWP) is a larger format that is preferable to the RFP. The following paragraphs will detail the main parts of that type of document.

The purpose of such documents is to provide the vendors with the specific information and requirements (discussed earlier) for the job estimates, including materials and labor.

The following format will assure the concise, complete work proposals that are essential for accurate bidding purposes.

For large, complex, or special work assignments, corporate management may require an extra layer of administration, namely a selection committee for the purposes of selecting the best vendor or contractor to be awarded the work.

Finally, be mindful of the bidding process and what it implies. An extremely low bid may be submitted by a vendor to attempt to obtain future business. Very low bids may not contain a reasonable amount of profit. This may indicate that in an attempt to get the present work assignment, the contractor may be looking for ways of cutting costs outside of specifications.

The following is a general outline for a DWP:

Section 1   General specifications
Section 2   Contracts and definitions
Section 3   General installation guidelines
Section 4   Work assignments
Section 5   Equipment and materials list
Section 6   Architectural and layout drawings
Section 7   Testing procedures

Section 8   Database updates and revisions

Section 9   Quality control checkpoint

Section 10 Corporate acceptance tests

Section 11 Vendor warranty

Figure 12-2 is an example of a detailed work proposal form.

Section 1 General Specifications

1. National and Local Compliance

    All work will be done in compliance with the national, state, and local codes (including, but not limited to, building, electrical, fire, and health).

2. Corporate Specifications

    All materials (cabling, electrical, electronic, etc.) must meet corporate specifications as outlined in this document.

3. Vendor Prior Work Experience

    The vendor awarded this contract shall have prior experience with this type of installation or work activity. Proof of the prior work shall be made available upon request.

4. Corporate Compliance

    All work shall be in compliance with the following corporate standards and publications: In this section, if the work to be performed is to meet existing corporate criteria, all references should be noted, and availability and access resources should be noted so that the contractors can review the various criteria. If no additional criteria are required, this section should be omitted.

5. Cleanup and Damage

    The contractors shall be responsible for cleanup in all areas in which work is performed, including reinstallation of all equipment disturbed during the work activity. The contractor is also required to replace ceilings and floor tiles, repair walls, floors, and so on damaged during work activity.

6. Service Interruption and Reconnection

    Vendors shall be responsible for the reconnection of any communication, telephone circuitry, and/or power disrupted during the work activity.

7. Time Target Schedule

    A schedule showing a time line listing all work by period divisions will be provided. Specific targets will be monitored by time/day acceptance by the project coordinator.

8. Access

    Details of how the vendor will gain access to the work area should be presented here. If the area is a secured area, plant security will also have to be notified as to who will be needing access and for how long.

REQUISITION #

REQUESTOR NAME	DEPT.	WK PH	WORK *o CODE	CURRENT DATE

ORDER NUMBER	JOB/PROJECT NUMBER

NEED DATE

TYPE OF ACTIVITY REQUESTED *1

DESCRIPTION OF PURCHASE

	ACTIVITY	PARTS REQ'D	PART #	QUANTITY	UNIT PRICING	TOTAL
ITEM 1						
2						
3						
4						
5						
6						
7						
TOTAL (ESTIMATE)............						

*o WORK CODE

701 = ENGINEERING
702 = TEST
703 = R & D
704 = SPECIAL PROJECT
705 = SUPPORT

ACTIVITY *1

501 = PRODUCT INSTALLATION
502 = SERVICE REQUEST
503 = PRODUCT & SERVICE REQUIREMENTS
504 = OUTSOURCE LABOR
505 = VENDOR

APPROVAL SECTION

MANAGEMENT LEVEL 1: _____ _____ _____ _____
                      NAME   PH   JOB CODE   AUTHORIZATION.
                                                    LEVEL

MANAGEMENT LEVEL 2: _____ _____ _____ _____
MANAGEMENT LEVEL 3: _____ _____ _____ _____

ACCOUNTING SECTION

IF CAPITAL PURCHASE: ( ) CAPITAL   COST _____ (REQUIRED)
                                    BUDGET _____ (BY DEPT./FUNCTION)

                     ( ) NEW   ( ) E/C
                     ( ) EXPENSE   COST _____
                                    BUDGET _____

                     ( ) OTHER _____ (REQUIRED IF CHECKED)

ACCOUNT APPROVAL

                     /BUDGET
          APPROVER/PLANNER _____ (NAME) EXT _____
          ACCT. NUMBER: _____
          CAPITAL PLAN/PROJECT: _____

( ) FIXED ASSETS: _____
( ) ACTUAL TOTAL COST OF REQUEST $ [____]   VERIFIED NO _____
                                               YES _____

                                   HIGHEST LEVEL
                                   OF MANAGEMENT APPROVAL: _____

SAFETY HAZARDS

PERSONNEL   ( )   _____
CHEMICAL   ( )   _____
ELECTRONIC ( )   _____
ELECTRICAL ( )   _____
FACILITY   ( )   _____   _____/_____
                                        NAME   DATE

Figure 12-1 Example of a RFP form.

Attaching Products Worksheet (Continued)

	Accessories	Part Number	Total Number	Comments
	Multiuse Communication Loop			
MCL-1	Type 1LS Loop Station Connector (LSC)	4760511		
MCL-2	Loop Wiring Concentrator (LWC)	6091077		
MCL-3	Component Housing	6091078		
MCL-4	Cable Bracket	6091042		
MCL-5	Patch Cable (8 feet)	8642551		
	Series/1			
S/1-1	MFA/422 Attachment Cable	8310553		
S/1-2	Y Assembly	8642549		
S/1-3	Twinaxial Y Assembly	8642550		
S/1-4	Twinaxial Straight Adapter	7362230		
S/1-5	Patch Cable (8 feet)	8642551		
S/1-6	Patch Cable (30 feet)	8642552		
S/1-7	Series/1 Feature #5790	- - - - - -		
S/1-8	Twinaxial Impedance Matching Device	6091070		
S/1-9	Twinaxial Terminator	6091068		
S/1-10	Twinaxial Direct Connect Cable	6091075		
	5080 Graphics			
5080-1	Red Coaxial Balun Assembly	8642546		
5080-2	Single Cableless Balun Assembly	6339082		
5080-3	Double Cableless Balun Assembly	6339083		
5080-4	Y Assembly	8642549		
	General Purpose Attachment			
Gen-1	General Purpose Attachment Cable	8310554		
Gen-2	Patch Cable (8 feet)	8642551		
Gen-3	Patch Cable (30 feet)	8642552		

Figure 12-2 Example of a DWP form.

9. **Work Impact**

While the work activity is in progress, there may be interruptions to the communications network, computing system, building power, lighting systems, and so on. All "impacted areas" need to be identified, and all affected departments must be notified well in advance so that they can adjust their work schedule.

10. **Labeling**

All cabling runs, conduits, and terminal blocks, and so on shall be labeled in compliance with all corporate and required standards.

Complete Order Summary Worksheet   (Part 1 of 4)

Cables: For installation and maintenance, order 15% additional cable.

Type	Part Number	Meters (feet)
1	4716748	
1 Plenum	4716749	
1 Outdoor	4716734	
2	4716739	
2 Plenum	4716738	
5	4716744	
6	4716743	
8	4716750	

Accessories:

Description	Part Number	Quantity
Cable Tester Kit (includes tester, case, data wrap plug, and batteries)	4760500	
Cable Tester (includes batteries)	4760501	
Twinaxial Test Accesories (includes twinaxial test adapter, twinaxial test terminator, and two twinaxial straight adapters	6339087	
Telephone Tester Attachment Kit	4760509	
Data Wrap Plug	4760507	

Equipment
Racks: Racks are not available from IBM. Order from your electrical supplier or contractor. Racks may not be a stock item, so allow enough lead time.

Type	Quantity
Open Rack	
Enclosed Rack	

Note: For large installations where extensive tester usage is anticipated order:
- One 8-foot patch cable
- Additional data wrap plugs

This will extend the life of the data test cable connector and also facilitate testing multiple offices from the wiring closet.

Figure 12-3

Complete Order Summary Worksheet   (Part 2 of 4)

Accessories: For installation and maintenance, order 10% additional accesoreis.
Order at least two additional surge suppressors of each type used.

Description	Part Number	Quantity
Data Connector *	8310574	
3-Pair Telephone Jack *	8310575	
3- or 4-Pair Telephone Jack *	8310551	
Type 1 Faceplate *	8310572	
Type 1 Faceplate for Japan *	6339094	
Type 2 Faceplate for 3-Pair Telephone Jack *	8310573	
Type 2 Faceplate for 3- or 4-Pair Telephone Jack *	6091025	
Type 2 Faceplate for 3- or 4-Pair Telephone Jack for Japan *	6339095	
Type 1W 87mm *	6091048	
Type 1W 80mm *	6091049	
Type 1S Surface Mt	4760486	
Type 2S Surface Mt for 3-Pair Telephone Jack	4760485	
Type 2S Surface Mt for 3- or 4-Pair Telephone Jack	6091029	
Distribution Panel	8642520	
Rack Ground Kit	4716804	
Surge Suppressor	4760469	
Cable Loc Chart	4716816	
Cable ID Label (8 sheets)	4716817	
Undercarpet Cable Connector Kit* (Note 3)	6339123	
Floor Monument (Note 3)	6339128	
Floor Monument Faceplate Kit (Note 3)	6339131	
Undercarpet Cable Wall Box (Note 3)	6339130	

Note:
  * Can only be ordered in multiples of 25
  3. Not available from IBM

Figure 12-4

Complete Order Summary Worksheet (Part 3 of 4)

Accessories: For installation and maintenance,
order 10% additional accessories.

Accessories used in more than one application

Description	Part Number	Quantity
Loop Wiring Concentrator (LCW)	6091077	
Cable Bracket	6091042	
Red Coaxial Balun Assembly	8642546	
Single Cableless Balun Assembly (note 3)	6339082	
Double Cableless Balun Assembly (note 3)	6339083	
Y Assembly	8642549	
Twinaxial Y Assembly	8642550	
Twinaxial Impedence Matching Device	6091070	
Twinaxial Direct Connect Cable	6091075	
Twinaxial Terminator	6091068	
Patch Cable (8 feet)	8642551	
Patch Cable (30 feet)	8642552	
General Purpose Attachment Cable (note 1)	8310554	
Coaxial Accessories		
Coaxial Patch Panel	4176801	
Yellow Coaxial Balun Assembly	8642544	
Single DPC Attachment Cable (8 feet)	6339073	
Single DPC Attachment Cable (20 feet)	6339074	
Double DPC Attachment Cable	6339075	
3299 Mounting Shelf	6217036	
Spare BNC Bulkhead Connector	(note 2)	
Twinaxial Accesssories		
Twinaxial Test Accessories Kit	6339087	
Finance Communication Loop Accessories		
Plug and Jack Assembly	8310552	
Y Assembly	8642549	
Store System Loop Accessories		
WE Type-404B Receptacle	(note 3)	

Notes:
1. Can be ordered for use as data wire test cable
2. Not available from IBM. Order Amphenol 31-220 or equivalent.
3. Not available from IBM.

Continued

Figure 12-5

Complete Order Summary Worksheet (Part 4 of 4)

Accessories: For installation and maintenance,
order 10% additional accessories.

Accessories used in more than one application

Description	Part Number	Quantity
Multiuse Communication Loop Accessories		
Type 1LS Loop Station Connector (LSC)	4760511	
Loop Wiring Concentrator (LWC)	6091077	
Component Housing	6091078	
Series/1 Accessories		
MFA/422 Attachment Cable	8310553	
Twinaxial Straight Adapter	7362230	
Series/1 Feature #5790	-------	

Figure 12-6

Rack Inventory Chart

Wiring closet number _____

Rack number _____

Date _____

Planner's initials _____

Instructions
Fill out a Rack Inventory Chart for each
equipment rack.

1. Enter the wiring closet location
    number, the equipment rack
    identification number, and the
    planner's initials.

2. Using the template for the
    Rack Inventory Chart that came
    with this manual, draw an outline
    of each component that will be
    installed in the rack.

3. The slots at the bottom of the
    distribution panel tempate are
    used only for the lowermost
    distribution panel in a rack.
    The slots indicate that there
    are 38.1 mm (1 - 1/2 in.)
    between that panel and the
    next unit in the rack.

4. Write the unit identification
    number on each component
    on the chart.

Example:

| 21 |
| 22 |
| 0010 |
| 0011 |
| 0012 |

Figure 12-7

Section 2 Definitions and Contracts

This section is for application-specific definitions for the common word usage.

2.1. The following are definition examples:

Company: This is the common name as referred through this document for the requesting company.

System Configuration Worksheet					
System		**Service Contact**		**Telephone**	
Attachment Description	Accessories in Work Area	Cable Runs from (Wall)	Cable & Cable Length	Cable Runs to (Panel)	Accessories on Distribution Rack
☐		☐		☐	
☐		☐		☐	
☐		☐		☐	
☐		☐		☐	
☐		☐		☐	
☐		☐		☐	
☐		☐		☐	
☐		☐		☐	
☐		☐		☐	
☐		☐		☐	

Suggested Accessory Abbreviations

GPA —	General Purpose Attachment Cable	RCB —	Red Coaxial Balum	LSC —	Loop Station Connector
MFA —	Multifunction Attachment Cable	SCB —	Single Cableless Balum	LWC —	Loop Wiring Concentrator
Y —	Assembly	DCB —	Double Cableless Balum	PJ —	Plug and Jack Assembly
TY —	Twinaxila Y Assembly	YCB —	Yellow Coaxial Balum	AD —	Adapter
IMD —	Impedance Matching Device	SDPC —	Single DPC Attachment Cable	P —	Patch Cable
TDC —	Twinaxial Direct Connect Cable	DDPC —	Double DPC Attachment	CPP —	Coaxial Patch Panel
				SS —	Surge Suppressor

Figure 12-8

Wiring Closet/Controller Room Worksheet							Building —————— Floor —————— Worksheet ——————	

Cable Routes Within a Single Building

	Wiring Closet or Controller			Cable Requirements				
Wiring Closet Location/ Floor	Room Location/ Floor	Number of Cables	Cable Length	Type 1	Type 1 P	Type 5	Faceplate Devices 1 1S 1W	
1								
2								
3								
4								
5								
6								
7								
8								
9								
10								
11								
12								
13								
14								
15								
	Totals							

Cable Routes Between Buildings

		Wiring Closet or Controller		Cable Requirements							
Wiring Closet Location/ Floor	Surge Suppressor Location/ Floor	Room Location/ Floor/ Building	Length of Indoor Cable in this Building	Type 1		Type 1 P		Length of Outdoor Cable	Type 1 Outdoor		Surge Sup- pressors
				No.	Total Feet	No.	Total Feet		No.	Total Feet	
1											
2											
3											
4											
	Totals										

Data Connectors ———————	Distribution Panels ———————	Rack Grounding Kit ———————
	Distribution Racks ———————	Cable Label Packages ———————

Figure 12-9

Cable1mm: Plenum cable 93 ohms fully connectorized with male-to-male connectors.

Cable 2: Ethernet thin cable.

Cable 3: Fiber-optic cable pair.

Outlet 2: IBM cable type 1 connectorized wall plate plus one RJ-11 type 3 phone jack.

Premise wiring system: Existing wiring and cable media systems.

2.2.   This section is for work-related contracts. Examples follow:

Projector coordinator: Lead person designated to coordinate all work activities and groups.

    Name: _____   Day Phone: _____

Support service contract: In the checklist of "affected groups" there should be a hardware and software contact as well as a system contact person. All phone numbers should be given and backup personnel should be listed.

    Hardware Support. Name: _____   Ph: _____
    System Support: Name: _____   Ph: _____

Section 3 General Installation Guidelines

This section is for installation specifications and requirements. If, for example, you are using this document to obtain an installation quote on voice, data, telco, ISDN, or fiber cabling, this section will assist the requestor to define such items as

1. correct cable type and impedance rating.
2. proper terminations for each cable group.
3. minimum and maximum cable length in work areas.
4. appropriate outlet fixtures for office, lab or other work areas.
5. proper labeling consistent with corporate and national standards.
6. definition of cable pathing. Reference premise cable runs and layout. Ensure that all runs are within cable trays where specified.
7. reference exhibits for all cable trays and cable runs.

If any component of the work must adhere to standards (i.e. corporate), this would be the area to refer to them by publication number or index resource.

Section 4 Work Assignments

In this section the actual work description will be given. The following is an example of two typical assignments.

Work Assignment 1

- Install IBM cabling system rack for use with IBM 3270 products on computer floor 3 location j4–18.
- Port wiring harness (coax) from control unit (3x) shall be connected directly to back of coax patch panel.
- Hardware required at computer floor:
  - 1 equipment rack
  - 1 distribution panel
  - 4 24 BNC female/female coaxial patch panels
  - 1 3299 multiplexer
  - 1 equipment shelf
  - 1 interconnection type 1, or 9 cable run between back of distribution panel to the office area
  - 10 cable guides
  - 20 interconnection cables
  - 10 jumpers
- Hardware required at office:
  - 1 type-1 face plate
  - 1 single cable coaxial balun assembly (which can be used instead of red balun assembly (93 to 150 ohm converter)
  - 1 coax cable

  Please note all required hardware and equipment lists are provided in the following section.
- Installation details are covered in Section 6 of this document.

Section 5 Equipment and Materials List

Description	Part Number	Quantity	Source	Comments
Equip. rack	8899xx–2	1	Vendor	
Equip. shelf	0008989	2	Vendor	
Distrib. panel	5555555	1	OEM	
3299-1 Multiplexer	0xxxx78	1	OEM	
Coax patch panel	9090.8	4	Vendor	Generic
Cable (IBM type 1)	CCCxxx	2000ft	OEM	
Cable connectors	909090	50	OEM	
Baluns	1m9898	6	OEM	
Coax (93 ohm)	NNNNN	400 ft	Vendor	Generic
Cable guides	poppp8	20	Generic	
Coax terminations	1787	50	Vendor	
Face plates	87877n	1	OEM	

SIGNALS FROM MODEM

Received data	Pin 3	Data from CPU.
Clear to send	Pin 5	Tells terminal that it may now place data on the transmit data line (pin 2).
Data set ready	Pin 6	Tells terminal modem is connected. Powered up and ready.
Signal ground	Pin 7	Common ground reference for all signal lines.
Protective ground	Pin 1	Safety or power line ground for equipment.
Received line sig. detector	Pin 8	Tells terminal that carrier is being received from computer modem.
Transmission sig. element timing	Pin 15	Clock signal from modem (used) only with synchronous modems).
RCVD. sig. element timing	Pin 17	Identical in function to pins 3 and 5. Except as they apply only to systems with full secondary channels implemented.
Secondary received data	Pin 16	
Secondary clear to send	Pin 13	
Secondary RCVD. line sig. detect	Pin 12	Tells terminal that carrier is present on secondary channel. Used for HD supervisor operation.
Signal quality cetector	Pin 21	Used by some modems which incorporate signal evaluating circuitry to advise terminal that present signal is poor and a high error rate is probable.
Data rate sig. selector	Pin 23	Used by modem/terminals with programmable data rate selection.

Pins receiving signals from terminal: 1,2,4,7,14,19,20,23,24.

SIGNALS FROM TERMINAL

Pin 2	Transmitted data	Data from terminal
Pin 4	Request to send	Tells modem that terminal wants to send data.
Pin 20	Data terminal ready	Tells modem that terminal is connected. Powered up and ready.
Pin 7	Signal ground	Common ground reference for all signal lines.
Pin 1	Protective ground	Safety or power line ground for equipment.
Pin 24	Transmit sig. element timing	Clock signal from terminal.
Pin 14	Secondary transmitted data	Identical in function to pin 2 except it applies only to systems with full secondary channel implemented.
Pin 19	Secondary request to send	Tells modem to turn on the secondary channel carrier. Used for HD supervisor operation.
Pin 23	Data rate signal selector	Used by modem/terminals with programmable data rate selection.

Pins receiving signals from modem: 1,3,5,6,7,8,12,13,15,16,17,21,23.

Section 6 Installation Specifications

Assemble IBM cabling system and install rack in location "A" as shown on layout 1 including orientation.

Once all mechanical hardware is assembled begin recabling process. Note: Below tiles marked "FF" is the controller port coax harness. Locate and connect to coax patch panels at bottom of equipment rack through cable access cut floor tile. All other connections and links will be performed by company personnel.

Section 7 Architectural and Layout Drawings

The following layouts are for work assignments 1 and 2.

Section 8 Testing and Debugging Procedures

The contractor is to perform all first- and second-level problem determinations on all vendor-installed equipment and services. Vendor shall provide all necessary test equipment and personnel to ensure that adequate testing and debugging is an internal part of any testing procedure.

Contractor is to ring out all single and multicable bundles once connected. All high-resistance and open runs will be corrected prior to release back to project coordinator.

Section 9 Database Updates and Revisions

After all moves, adds, and changes are completed, the project coordinator shall be responsible for getting the database updated to reflect current conditions. This task can be delegated to database management.

Section 10 Quality Control Checkpoint

A person from corporate management, preferably from quality control, shall be designated to ensure that all major tasks are completed in accordance with existing guidelines.

A checklist should be provided that begins with inventory, work activities, and so forth, including the final step of database updating.

Section 11 Corporate Acceptance Testing

Vendor shall notify project coordinator of work assignment competition. The appropriate corporate function will then perform mandatory acceptance testing. Any problems noted during these acceptance tests will be reported to vendor for correction. Time is of the essence. All corrections are to be made within an accepted time limit as prescribed by corporate directive.

Upon satisfactory acceptance testing of a work assignment, a corporate acceptance letter shall be issued.

Section 12 Vendor Warranty

Vendor shall warranty all materials and labor for a minimum of one year from date of corporate acceptance.

Summary

The proposal for bid submissions should be written only after a detailed study of the project, and then by a person who is committed to detail. Finally, the proposal should be reviewed by the department "nitpicker" before being submitted to contractors.

Questions

1. Why is it important that the specifications for the call bids be very detailed?
2. When should a low bid be suspect?
3. Should a corporation expect every contractor to have experience in the area of a bid?
4. Who is usually responsible for testing troubleshooting a cabling installation?
5. Who performs the final acceptance test?
6. When does the vendor's guarantee begin?

Glossary of Terms

Access Line The connection between the subscriber's facility and the public network.

ALOHA An experimental LAN of the packet-switching type utilized by the University of Hawaii.

AM Modulation One of the methods of inserting information into a carrier. In this case, the amplitude of the carrier is varied at the rate of the carrier signal.

Amplifier An electronic device used to boost (amplify) a signal.

Amplitude Distortion The unwanted change of the amplitude of a signal. This distortion is sometimes called *noise*.

Analog A format that uses continuous physical variables such as voltage amplitude or frequency variations to transmit information.

Application Layer The seventh and highest level in the OSI model that contains the user application programs.

Aramid^{TM} Yarn Strength member element used in Siecor cable to provide support and additional protection of the fiber bundles. Kelvar is a particular brand of Aramid^{TM} yarn.

Architecture The structuring of a LAN or the operation of a computer.

Armoring Additional protection between jacketing layers to provide protection against severe outdoor environments. Armoring is usually made of plastic-coated steel and may be corrugated for flexibility.

ASCII The American Standard Code for Information Interchange, an eight-bit code used for information exchange. Seven bits represent the characters, and the eighth bit is used for parity.

Asynchronous Transmission A transmission method in which the time interval between characters may be of unequal length. Asynchronous transmission is sometimes called start-stop transmission.

Attenuation The decrease in magnitude of power of a signal in transmission media between points. A term used for expressing the total loss in an optical fiber consisting of the ratio of light output to light input. Attenuation is usually measured in decibels per kilometer at a specific wavelength.

AT&T American Telephone & Telegraph, Inc.

Baluns Impedance-matching devices comprised of a transformer that connects a balanced line (twisted pair) to an unbalanced line such as a coaxial cable.

Bandwidth Measure of the information-carrying capacity of an optical fiber, normalized to a unit of MHz/km. This term is used to specify capacity of multimode fibers only. The term dispersion is used for single-mode fibers.

Baseband A frequency band occupied by a single signal in its original unmodified form.

Baseband Transmission A transmission method that uses low frequency starting at zero hertz and carries only one transmission at a time. This method utilizes no carrier.

Baud The rate at which signals are transmitted. A bit rate of 2000 is a baud rate of 2000.

Bend Radius The amount that a fiber can bend before the risk of breakage or increase in attenuation. Also can refer to the maximum bend of a coaxial cable.

BICONIC A type of connector for a fiber-optic cable.

Binary The two states of a digital system comprised of ones and zeros or signals and no signals.

BNC A type of connector with a bayonet-type locking mechanism.

Breadboard A thin plastic board full of holes on which components are mounted.

Breakout Box A device used for testing a circuit or cable.

Breakout Cable *See Fan-out cable.*

Broadband In data transmission, transmission facilities capable of handling frequencies greater than those of high-quality voice communications. The higher frequency allows the carrying of several simultaneous channels. Broadband infers the use of a carrier signal rather than direct signal transmission.

Buffer A temporary storage device made up of logic circuits. Such a device is used to isolate one system from another in which the data flow, sequence of events, or impedance are different.

Buffering (1) A protective coating extruding directly on the fiber coating to protect it from the environment; (2) extruding a tube around the coated fiber to allow isolation of the fiber from stress on the cable.

Bundle Many individual fibers or other cables within a single jacket or buffer tube. Also a group of buffered fibers distinguished in some fashion from another group in the same cable core.

BUS A common set of cables for a multiuser computer system. A coaxial cable-type of communication cable that usually has a single center conductor surrounded by an insulation material and a braided shield.

Cabinet An enclosure for rack-mounted equipment.

Cable Bend Radius During installation, the term infers that the cable is experiencing a tensile load. Free bend infers a lower allowable bend radius since it is at a condition of no load.

Cable Through The capacity of an information system that allows multiple workstations to attach to a single cable.

CAD Computer-aided drafting.

Call A request for a connection to the system.

Camp-On A PBX or LAN operation that allows a caller (user) to wait on line if the called station is busy.

C Connector A type of bayonet locking connector for coaxial cable.

Central Member The center component of the cable. It serves as a strength member or an antibuckling element to resist temperature-induced stress. The central member is comprised of steel, fiberglass, or glass-reinforced plastic.

Character A symbol such as a number, letter, punctuation, or control function.

Cladding The coating on the fibers of a fiber-optic cable.

CO Central office.

Code The rules that specify the way that data are to be presented and read.

Control Unit The unit that directs traffic between a host computer and the I/O devices.

Core The central region of an optical fiber through which light is transmitted.

CPC Customer premises communication.

Crimp To secure the buffer tubing to a fiber cable to the tabs on a connector, or to secure the shield of a coaxial cable to a connector.

Cross-Talk The transfer of an unwanted signal from one circuit to another.

Current A movement of electrons through a circuit. The force that carries the energy through the circuit. The unit of current is the ampere (A).

Database A large collection of information in computer files.

Data Communications The process of transferring information from one data processing device to another. The transfer is usually in digital form.

dB A comparison of signal levels based on the base 10 logarithm functions. The comparison can be between power levels, voltage levels, or current levels. A 50% power loss is a $-3dB$ loss.

DEC Digital Equipment Company—one of the leading manufacturers of microcomputer-related hardware and software.

Device An input/output unit such as a printer, PC, or other workstation.

Diagnostics Procedures used to test and troubleshoot a system.

Dielectric The insulation material between the center conductor of a coaxial cable and the shield. Nonmagnetic and, therefore, nonconductive to electrical current. Glass fibers are considered dielectric. A dielectric cable or material contains no metal.

Disk An electronic storage device for digital information.

Dispersion The cause of bandwidth limitations in a fiber. Dispersion causes a widening of input pulses along the length of the fiber.

Distortion Any unwanted deviation of the information signal. Distortion may be a change of amplitude, phase, frequency, or shape of a signal.

Down Time The time when all or part of a system is out of operation and not available to the user.

Drop The connection on a circuit, a cable from a distribution panel to a face plate.

Echo The return of a signal on a line usually due to an impedance mismatch.

EIA Electronics Industries Association An electronics industry association and a standards association that publishes test procedures.

Fan-out Multifiber cable constructed in the tight buffering design. Designed to ease of connectorization and rugged applications for intra- or interbuilding requirements.

Fault An open break or an intentional short in a circuit.

FCC The Federal Communications Commission. A U.S. government agency that has the responsibility of setting the standards for electronic transmission and their enforcement.

Ferrule A mechanical fixture, generally a rigid tube, used to confine and align the stripped end of a fiber.

Fiber Thin filaments of glass. An optical waveguide consisting of a core and a cladding which is capable of carrying information in the form of light.

Fiber Optics Light transmission through optical fiber for communication.

File A collection of data records.

FM Modulation Frequency modulation of a carrier. A method whereby the carrier frequency is shifted in rhythm with the information frequency.

FOTP Fiber-Optic Transmission System.

Fusion The actual operation of joining fibers together by fusion or heat.

Fusion Splicing Splicing of fiber cable with heat.

Graded index Fiber design in which the refractive index of the core is lower toward the outside of the fiber core and increases toward the center of the core.

Grommet Plastic or rubber edging used around cable and pigtail entrance to holes such as termination centers.

Handshaking Part of the communication protocol in data transfer.

Hertz (Hz) A measure of frequency in cycles per second. For example, 100 cycles per second equals 100 Hz.

Hexadecimal A digital numbering system used for coding, consisting of 16 bits. The bits are numbered 0 through 9 followed by A, B, C, D, E, and F.

Host Computer The controlling or central computer in a data communication network.

IDF Intermediation distribution frame.

IEEE Institute of Electrical and Electronic Engineering.

Index of Refraction The ratio of light velocity in a vacuum to its velocity in a given transmission medium.

Input Device A device in a data processing system from which data may be entered into the system.

Interface The connection between equipment or the equipment that interfaces one system to another.

ISDN Integrated Services Digital Network A CCITT standard that is primarily concerned with the control of voice and data.

ISO International Standards Organization An international standards organization that has developed an OSI communication model for data communication protocol. The OSI model is widely accepted in the international community.

Kilometer One thousand meters, or 3281 feet. The kilometer is the unit of measurement distance in most of the world.

KPSI A unit of tensile strength expressed in thousands of pounds per square inch.

LAN See *local area network*.

Laser Light amplification by stimulated emission of radiation. A device which produces coherent light with a narrow range of wavelengths.

Leased Line A telephone line reserved for the leasing party.

LED Light-emitting diode. A device that is used as a readout matrix, as an indicating light, and as light transfer to electrical signal in a fiber-optic system. Also a device used in a transmitter to convert information from electric to optical energy.

Link A fiber-optic cable with connectors attached to a transmitter (source) and receiver (detector).

LLOPE Linear low-density polyethylene jacketing.

Local Area Network (LAN) A geographically limited communications network intended for the local transporatation of data, voice, and video.

Local Exchange The central or local exchange where the subscribers lines terminate.

Loop A unidirectional closed signal path connecting input/output devices.

Mainframe A mainframe computer. The host computer.

Mbps A million bits per second.

MDF Main distribution frame.

MDPE Medium-density polyethylene jacketing.

Mechanical Splicing The joining two fibers together by mechanical means to enable a continuous signal.

Megahertz (MHz) A unit of frequency that is equal to 1 million cycles per second.

Metric Prefixes A group of symbols that are used to indicate powers of 10. For example, 1000 or 10^3 is written 1 k.

MICON Systems, Inc. A leading supplier of data communications equipment.

Micron (μm) Another term for micrometer. One-millionth of a meter.

Microprocessor A single chip which contains the processor of a computer.

Minibundle Cable Siecor cable in which the buffer tube contains three or more fibers.

Mode A term used to describe a light path through a fiber, as in multimode or single mode.

Modem A device that converts serial data transmission to parallel data transmission. Also a device that converts digital information to analog or analog to digital.

Modulation Coding of information on to the carrier frequency. This includes amplitude, frequency, or phase modulation techniques.

Monitor The TV-like screen of a workstation.

Multichannel Cable A cable on which more than one channel of information can be transported. Can be base band or wide band.

Multifiber Cable An optical cable that contains two or more fibers, each of which provides a separate information channel.

Multimode Fiber An optical waveguide in which light travels in multiple modes.

Multiplex To put two or more signals into a single channel.

Multipoint Line A communication line or circuit that interconnects several stations or terminals.

MUX A multiplexing system used to transmit more than one signal over a transmission media.

N-Type Connector A threaded connector for connecting coaxial cable.

Nanometer A unit of measurement equal to one-billionth of a meter.

NEC National Electrical Code Defines the building restrictions for cable, such as the flammability requirements.

Network The interconnection of computer systems: PCs, workstations, and so on.

Noise Unwanted electrical signals interfere with data transmission. Noise can be internal or external to the system.

Numerical Aperture The number that expresses the light-gathering power of a fiber.

Octal An eight-bit numbering system consisting of numerals 0 through 7 that can be used for a digital code.

Optical Fiber Glass fiber strands in a fiber-optic cable along which the light energy is transported.

OTDR Optical time domain reflectometer Used to measure light fiber attenuation in fiber and across a connector.

Output Device A device in a data processing system from which data may be received from the system.

PBX Private Branch Exchange A user's branch exchange.

PE Abbreviation used to denote polyethylene. A type of plastic material used in manufacture of plastic material used for cable jacketing.

Perfusing Fusing with a low current to clean the fiber ends as a procedure to fusion splicing.

Phase Modulation A type signal modulation where the phase of the carrier signal is shifted by the modulation signal.

Pigtail Fiber-optic cable that has connectors installed on one end. See *Cable assembly*.

Pin Diode A device used to convert the optical signals to electrical signals in a receiver.

Plenum An air duct inside a building through which cable can be housed.

Plenum Cable A cable that has been UL approved as having high heat resistance, low toxic, and smoke-producing properties that allows it to be installed in heating air ducts.

Point-to-Point Connection A communication circuit that "links" two points.

Port A computer interface point, usually a connection where information can be inputted or outputted to communicate with another device.

Protocol The procedures of operation that are followed by a computer operating system.

PTSS Passive Transmission Subsystem.

PUR Polyurethane material used in the manufacture of a type of jacketing material.

PVC Polyvinyl chloride material used in the manufacture of a type of jacketing material.

Receiver An electronic package which converts optical signals to electrical signals. Can also be a device to receive electromagnetic radio waves and amplify them for a useful purpose.

Refractive Index See *Index of refraction.*

Repeater A device which consists of a transmitter and a receiver or transceiver, used to regenerate a signal to increase the system transfer range.

RING A type of connection for a computer system. A network in which a unidirectional signal path is a closed loop.

Riser Application Used for indoor cables that pass between floors, it is normally a vertical shaft or space.

Rise Time The measurement of the time it takes a pulse to rise from 10% of maximum value to 90% of maximum value.

RS-232 An EIA recommended standard for computer cable terminal connection that specifies the voltages and electrical character of the pin signals. The 25-pin connection is the most used cable connection.

Scattering A property of glass which causes light to deflect from the fiber and contribute to signal loss.

Serial Transmission A transmission where the bits are transmitted one at a time and in sequence.

Signal-to-Noise Ratio A ratio comparison of the desired signal to unwanted signals on the line or in the system.

Single-Mode Fiber An optical waveguide (or fiber) in which the signal travels in one ''mode.'' The source is usually a light-emitting diode (LED).

Software A computer program or computer programs that controls the operation of the logic circuits to perform certain functions.

Source The means used to convert an electrical information-carrying signal to a corresponding optical signal for the transmission by fiber. The source is usually an LED or laser.

Splice Closure A container used to organize and protect spliced trays.

Splice Tray A container used to organize and protect spliced fibers.

Splicing The permanent joining of fiber ends to identical or similar fibers, without the use of a connector. See also *Fusion splicing* and *Mechanical splicing.*

Star Coupler optical component that allows emulation of a bus topology in fiber-optic systems.

Station Any setup to perform a computer operation, for example, a PC station.

Step-Index Fiber Optical fiber that has an abrupt (step) change in its refractive index, due to a core and cladding that have different indices or refraction. Typically used for single-mode operation.

Tap A connection to the main transmission line in a LAN.

Telco An abbreviation for telephone company.

Teleprocessing The handling of data processing information utilizing communication equipment.

Telex A network of printers connected over an international public network.

Terminal Block A terminal strip with connections through and from which cables can connect.

Termination The connecting of a connector at the end of a cable or the termination of the cable in its characteristic impedance.

Tight Buffer Type of cable construction whereby each fiber is tightly buffered by a protective thermoplastic coating to a diameter of 900 microns, resulting in high tensile strength, durability, ease of handling, and ease of connection.

TNC A threaded type of connector used on coaxial cable.

TOKEN Ring A type of connection for a LAN computer system in which a code (token) is passed around the ring to a certain station. The station with the token can receive the information from the ring.

Training Time The time necessary to train an operator or user to operate a piece of equipment.

Transmitter An electronic package which converts electrical signals to optical signals or electrical signals to electromagnetic waves.

Twin-axial Cable A shielded cable consisting of one or more twisted-pair cables.

Twisted-Pair Wiring Two wires twisted to cancel out electromagnetic induction.

UL Underwriters Laboratories, Inc. A U.S.-based company that tests the standards of products for safety and reliability. Manufacturers pay to have their equipment tested to acquire a UL stamp of approval.

Up Time The time that a system is in normal operation.

Voltage The measure of electromotive force that forces current through a circuit.

Wavelength The distance between two crest of an electromagnetic or light signal.

Work Area An area in which personnel operate data processing equipment.

Workstations Areas in which there are data processing equipment such as terminals, printers, and so on.

X_C Capacitive reactance. The reactance or opposition that a capacitor offers to an ac current flow.

X_L Inductive reactance. The reactance or opposition that an inductor offers to an ac current.

TABLE G-1 Powers of 10, Prefixes, and Symbols

Powers of ten	Number	Prefix	Symbol
10^{12}	1,000,000,000,000	tera	T
10^{11}	100,000,000,000		
10^{10}	10,000,000,000		
10^{9}	1,000,000,000	giga	G
10^{8}	100,000,000		
10^{7}	10,000,000		
10^{6}	1,000,000	mega	M
10^{5}	100,000		
10^{4}	10,000		
10^{3}	1,000	kilo	K
10^{2}	100	hecto	h
10^{1}	10		
10^{0}	1	deka	dk
10^{-1}	.1	deci	d
10^{-2}	.01	centi	c
10^{-3}	.001	milli	m
10^{-4}	.000 1		
10^{-5}	.000 01		
10^{-6}	.000 001	micro	μ
10^{-7}	.000 000 1		
10^{-8}	.000 000 01		
10^{-9}	.000 000 001	nano	n
10^{-10}	.000 000 000 1		
10^{-11}	.000 000 000 01		
10^{-12}	.000 000 000 001	pico	p

Table G-2 Electrical and Electronic Properties Designated by Greek Letters

Name	Capital	Lower case	Used to designate
Alpha	A	α	area, angles, coefficients
Beta	B	β	angles, coefficients, flux density
Gamma	Γ	γ	specific gravity, conductivity
Delta	Δ	δ	density, variation
Epsilon	E	ε	base of natural logarithms
Zeta	Z	ζ	coefficients, coordinates, impedance
Eta	H	η	efficiency, hysteresis coefficient
Theta	Θ	θ	phase angle, temperature
Iota	I	ι	
Kappa	K	κ	dielectric constant, susceptibility
Lambda	Λ	λ	wavelength
Mu	M	μ	amplification factor, micro, permeability
Nu	N	ν	reluctivity
Xi	Ξ	ξ	
Omicron	O	ο	
Pi	Π	π	3.1416
Rho	P	ρ	resistivity
Sigma	Σ	σ	summation
Tau	T	τ	time constant
Upsilon	Y	υ	
Phi	Φ	φ	angles, magnetic flux
Chi	X	χ	
Psi	Ψ	ψ	dielectric flux, phase difference
Omega	Ω	ω	ohms, angular velocity

Table G-3 Abbreviations, letters and symbols used in electronics
This list represents the most common abbreviations used in electronics.

alternating current	ac	megacycle	mHz
ampere	amp	megohm	MΩ
antenna	ant.	microampere	μa
antilogarithm	antilog	microfarad	μF
audio frequency	af	microvolt	μV
base of natural logarithms	ε	microvolts per meter	μv/meter
centimeter	cm	milliampere	mA
continuous waves	c-w	millihenry	mH
cycles per second	Hertz	millivolt	mV
decibel	dB	microvolts per meter	mV/meter
direct current	dc	milliwatt	mW
electromotive force	emf	ohm	Ω
electron volt	ev	picofarad	pF
feet	ft	power factor	PF
frequency	f	radio frequency	rf
inch	in.	revolutions per	
intermediate		minute	rpm
frequency	i-f	revolutions per	
kilocycle per second	kHz	second	rps
kilohm	kΩ	ultra high frequency	uhf
kilowatt	kW	very high frequency	vhf
logarithm	log	very low frequency	vlf
		yard	yd

References

BOWER, RICHARD, *Interoperability,*Foxborough, MA: Rascal Inter-Lan, 1988.

BOYD, WALDO T., *Fiber Optics Communication, Experiments, and Projects,* Indianapolis, IN: Howard Sams and Co., Inc., 1982.

CHORAFAS, DIMITRIS N., *Telephony, Today and Tomorrow,* Englewood Cliffs, NJ: Prentice-Hall, Inc., 1984.

CORTADA, JAMES W., *Managing DP Hardware, Capacity Planning, Cost Justification, Availability, and Energy Management,* Englewood Cliffs, NJ: Prentice-Hall, Inc., 1983.

DURR MICHAEL, and GIBBS, MARK, *Networking Personal Computers,* Carmel NI: Que Corp, 1989.

EDUCATIONAL SERVICES DEVELOPMENT, *Ethernet Fundamentals,* Nashua, NH: Digital Equipment, Inc., 1988.

FRIEND, GEORGE E., et al, *Understanding Data Communications,* Dallas: Texas Instruments, Inc., 1984.

GREEN, JAMES HARRY, *Telecommunications,* Homewood, IL: Dow Jones-Irvin, 1986.

GROSMAN, LAWRENCE, *Managers' Guide to the New Telecommunication Network,* Boston, MA: Artech House, 1988.

HANKCOCK, BILL, *Network Concepts and Architectures,* Wellesley, MA: QED Information Sciences, Inc., 1988.

HEEDMAN, ROBERT, *Telecommunication Management Planning,* Blue Ridge Sunset, PA: Tab Books, 1987.

HELD, GILBERT, *Data Communications Testing and Troubleshooting*, Indianapolis, IN: Howard W. Sams & Company, 1988.

HOUSEL, DARDEN, *Introduction to Telecommunications, the Business Perspective*, Cincinnati, OH: South-Western Publishing, Co., 1988.

IBM REGIONAL TELECOMMUNICATIONS FUNCTION, *The Western Regional Cable Standards*, San Jose, CA: IBM, 1989.

LOMMIS, MARY E. S., *Data Communications*, Englewood Cliffs, NJ: Prentice-Hall, Inc., 1983.

MARTIN, JAMES, *Future Developments in Telecommunications* (2nd Edition), Englewood Cliffs, NJ: Prentice-Hall, Inc., 1977.

MARTIN, JAMES, *Local Area Networks*, Englewood Cliffs, NJ: Prentice-Hall, Inc., 1988.

MAYNE, ALAN J., *Linked Local Area Networks*, New York, NY: John Wiley & Sons, 1986.

NATIONAL FIRE PROTECTION ASSOCIATION, *National Electric Code*, Quincy, MA, 1989.

PASAHOW, EDWARD J., *Microcomputer Interfacing*, New York, NY: McGraw-Hill Book Company, 1981.

REYNOLDS, GEORGE W. and RIECKS, DONALD, *Introduction to Business Telecommunications*, Columbus, OH, Merrill Publishing Company, 1988.

SCHUBERT, WERNER, *Communication Cables and Transmission System*, Berton, MI: Siemens, 1976.

SCHWADERER, DAVID W., *IBM's Local Area Networks: Power Networking and System Connectivity*, New York, NY: Van Nostrand Reinhold, 1989.

SCHWEBER, WILLIAM L., *Data Communication*, New York, NY: McGraw-Hill Book Company, 1988.

SENIOR, JOHN J., JR., *Optical Fiber Communication Systems*, Englewood Cliffs, NJ: Prentice-Hall International, Inc., 1985.

SHARMA, ROSHAR, *Network Systems, Modeling and Design*, T.de Sorisa School, IN, 1982.

SHERMAN, BARRY L., *Telecommunication Management, The Broadcast and Cable Industries*, New York, NY: McGraw-Hill, Inc., 1987.

STALLINGS, WILLIAM, *Data and Computer Communications*, New York, NY: Macmillian Publishing Company, 1985.

STANLEY, RICHARD A., *Data Communications and Networks*, Benton Harbor, MI: Heath Company, 1986.

STERLING, DONALD J., *Technician's Guide to Fiber Optics*, Albany, NY: Delmar Publishing, Inc., 1987.

WEINBERG, GERALD M., *Rethinking Systems Analysis & Design*, New York, NY: Dorset House Publishing, 1988.

WIENBERG, GERALD M., *Rethinking Systems Analysis & Design*, New York, NY: Dorset House Publishing, 1988.

WOODWARD, JEFF, *The ABC's of Novell NetWare*, San Francisco, CA: Sybex, 1989.

Periodicals

The following is a sampling of periodicals for the Telecommunication professional. Most are free for persons responsible for the purchasing of telecommunication components, materials, and systems.

Circuit Design, PMSI, 1790 Hembree Road, Alpharetta, GA 30201

Connect, The Journal for Network Computing, 3 Com Corp, 3165 Kifer Road, Santa Clara, CA 95052

High Performance Systems, CMP Publications, Inc., 600 Community Drive, Manhasset, NY 11030

JEE Journal of Electronic Engineering, 400 Madison Avenue, New York, NY 10017

PC Computing, America's Computing Magazine, Ziff Communications Company, P.O. Box 50253, Boulder CO 80321–0253

PC Magazine, P.O. Box 54093, Boulder, CO 80322

Technical Support, Technical Support, Inc., 4811 S. 76th Street, Suite 210, Milwaukee, WI 53220

Vendor Information

The following list of vendors is a small sampling of the thousands of manufacturers and distributors that handle Telecommunication products and materials. The authors suggest, for further reference the *Telephone Industry Directory* by Phillips' Publishing Inc., 7811 Montrose Road, Potomac, Maryland, 20854, (800)722–9120. This directory contains thousands of vendor addresses and phone numbers.

Allied Electronics-Sub Digitech, Inc.
(Lineman's tools, test equipment and wire and cable)
401 East 8th Street
Fort Worth, TX 76102
(817)336–5401

Amdahl Communications Inc.
(Data network controllers, data sets, data test equipment, data transmission equipment, multiplexers, and repeaters)
2330 Millrace Court
Mississauga, Ontario L5N IW2
Canada

American Cable Corp
(Coaxial, plastic insulated jacked, plenum, plug/connector-ended, and switchboard cable; key telephone connectors and modular retractable cords)
711 Cooper Street
Beverly, MA 08010
(416)821–9900

Anixter Brothers Inc.
(Adapters, wiring systems, products for voice, video, data, and power application. Provides manufacturing and refurbishment of telecommunication products; outside plant materials for telephone construction; customer premises, transmission, and central office equipment; earth station satellites; and local area network equipment. Distributes transmission and plant equipment products for AT&T, Belden, Cablec, BICC, Brand-Rex, IBM, Keptel, Porta, Raychem, Rome Siecor, Teleco Systems, Thermo Electric, 3M, and TRW.)
Skokie, Ill USA
(708)677-9480

AMP Inc.
(Cable assemblies and connectors)
P.O. Box 3608
Harrisburg, PA 17105
(717)564-0100

AMP Products Corp
(Interconnection products, including closures, fiber optics mechanical strippers, cross-connect systems, and under carpet wiring and accessories)
450 West Swedesford Road
Berwyn, PA 19312
(215)647-1000

AT&T Network Systems
(Manufactures network switching, transmission, and outside plant products)
475 South Street
Morristown, NJ 07960

AT&T Technology Systems
(Fiber optic cable, connectorized cable, circuit boards, connectors, converters, equalizers, multiplexers, fiber optic transmission systems, power supplies, and wire)
One Oak Way
Berkeley Heights, NJ 07922
(201)771-2000

Belden Electronic Wire and Cable
(Cables and assemblies, connectors, and data links)
2200 South U.S. 27
Richmond, IN 47375
(317)983-5200

Data Products International
(IBM-compatible telephone company management software that runs on IBM and handles accounting, capital credits, commercial billing, general plant, inventory, and toll rating applications)

P.O. Box 1176
Sugar Land, TX 77487
(713)491-7200

International Business Machines, Inc. IBM™
(Main frames, PCs, premise wiring, software, local area network installation, system servicing, wiring, and cabling hardware)
System Product Department
One Culver Road
Dayton, NJ 08810
1-800-IBM-2468

Nevada Western, Inc.
(Wire management systems and local area networking products, local area networking design and installation)
615 North Tassman Avenue
Sunnyvale, CA 94089
(408)734-2700

Index

A

AC voltage, 3
Access method, 96
Access time, 227
Air conditioning, 116
AM modulation, 227
Amplifier, 227
Amplitude modulation, 227
Analog signals, 13
Analog voltage, 3
Application layer, 227
Architecture, 227
Architecture drawings, 225
Armaid yarn, 227
Armoring, 228
ASCII, 228
Attenuation of signal, 17, 228
AT&T, 228

B

Balun, 41, 228
Bandwidth, 228
Bandwidth length product, 78
Barrel-type connectors, 52
Baseband transmission, 228
Baseboard, 143
Baud rate, 12, 228
Bend radius, 228
Biconic fiber cable connectors, 84, 228
Bid proposal forms, 210
Bid proposal:
development time, 209
research, 208
response time, 209
specifications, 211
Bidding plan, 111
Binary, 228
BNC, 228
Bonding, 115
Breadboard, 228
Breakout box, 228
Breakout cable, 228
Broadband, 229
Buffering, 77, 229
Building cable summary, 54
Bundle, 229
Bus network, 96, 229
Byte, 12

C

Cable compatibility, 37
Cable drop, 145
Cable hardware, 147
Cable impedance, 50, 52
Cable installation:
baseboard, 143
cable drop, 145
cable strategy, 123
coax, 150
copper, 135
direct burial, 146
fiber optic, 135

Cable installation (*cont.*)
  grounding, 155
  hardware, 147
  molding strip, 144
  overhead, 137
  riser, 144
  rules, 135
  tunnel, 145
  under floor, 139
  wiring plan, 131
Cable labeling standard, 177
Cable panel guide, 154
Cable pulling loop, 154
Cable requirement, 111
Cable termination:
  coaxial, 50
  fiber, 81, 85
  twisted-pair, 31
Cable through, 229
CAD blueprint, 109
CAD system, 121, 229
Call, 229
Camp on, 229
Capacitance, 8
Capacitive reactance, 8
Central member, 229
Character, 229
Characteristics impedance, 11
Chilled water requirements, 115
Cladding, 74, 229
Coaxial cable connectors:
  BNC, 60
  comparison, 63
  crimp type, 60
  summary, 53, 54
  three piece, 60
  TNC, 60
Coaxial cable type: RG 6, RG 8, RG 11,
    RG 58, RG 59, RG 62
Coaxial cables:
  advantages, 68
  applications, 68, 70
  assembly, 62
  characteristics, 50
  connectors, 60
  construction, 49
  disadvantages, 69
  grounding, 64

impedance, 50
installation, 155
shielding, 50
termination, 53
Coax/twisted pair cabling system, 153
Code, 229
Communication network, 95
Communication Society of America
    (CSA), 18
Connection strategy:
  two-point, 132
  three-point, 133
  4-point, 134
Continuity testing, 161
Contractor considerations, 209
Contractor response time, 209
Contracts, 219
Controlled access room, 146
Control unit, 230
Copper cable, 135
Core, 230
Corporate acceptance test, 225
Cost factors, 204
CPU, 230
Crimp, 230
Crimping tools, 65
Cross talk, 8, 17, 25, 230
Current, 3, 230
Custom-assembled cable, 169

D

Data terminal logs, 179
Database, 108, 230
Database development, 204
Database management system, 186
Database manager's responsibility, 187
Database revisions, 225
Database sample, 187
Database tracking system, 128, 185
Database updates, 225
dB, defined, 230
dB losses, 173
DC voltage, 2
Debugging, 225
DEC, 230
Decibel, 18
Detail work proposal (DWP), 200, 214

Device, defined, 230
Device ownership, 116
Diagnostic, 230
Dielectric, defined, 230
Dielectric of a cable, 49
Direct burial cable, 146
Disk, 230
Dispersion, 230
Distortion, 230
Distribution cable, 311
Distribution frame, 36
Documentation, 120
  blueprint, 178
  distribution log, 178
  labeling, 177
  responsibility, 127
  work area inventory sheets, 179
Down time, 230
Drop, 230

E

Echo, 230
EIA, 230
EIA RS-232 connector, 152
Electrical characteristics of cable
  coaxial, 50
  twisted pair, 37
Electrical properties:
  cable, 239
  table, 12
Electromagnetic interference, 116
Electromotive force, 2
Electronic control, 122
Energy source, 1
Energy transfer medium, 1
Environmental concerns, 115
Equipment list, 223
Ethernet network, 51

F

Fan out, 231
FCC, 231
Fiber, 231
Fiber optic cable:
  advantages, 80
  application, 91
  attenuation vs. length, 80

attenuatuion vs. wavelength, 81
bandwidth vs. length, 81
buffering, 77
characteristics, 77
construction, 76
figure of merit, 78
pulse spreading, 79
splicing, 81
strength member, 77
summary, 82
termination, 81, 84
theory of operation, 72
types, 79
wire strippers, 87
Fiber optic cable types:
  carpet, 79
  ribbon, 79
Fiber optics:
  security, 73
  theory of operation, 72
Field description:
  link and control panel, 199
  master panel, 189
  room panel, 194
Figure of merit, fiber cable, 78
File-based tracking system, 184
Floor cells, 141
Flux lines, 8
FM modulation, 231
FOTP, 231
Frequency division multiplexing (FDM), 29
Fusion splicing, 231

G

General installation guidelines, 222
General specifications, 211
Graded-index fibers, 74, 231
Gromment, 231
Ground, 15
Ground loop, 27
Ground potential, 1
Ground symbols, 15
Grounding, 15, 64, 115, 155
  earth, 16
  shield, 26
  twisted pair, 37
Grounding electrode, 65

H

Handshaking, 231
Handwritten logs, 179
Hertz, 231
Hexidecimal, 231
Hierarchical network, 99
Host computer, 231
Hot host service, 117
HVAC system, 121
Hybrid network, 99

I

IDF (intermediate distribution frame),
  231
IEEE, 231
Impedance, 9
Impedance matching, 11
Induced voltage, 7
Inductance, 7
Inductive amplifier, 167
Inductive reactance, 7
Input device, 232
Insulation:
  defined, 18
  temperature range, 22
  types, 19
Interface, 232
Intermediate distribution frame, 36
International Standards Organization
  (ISO), 102
Inventory, 118
ISDN (Integrated Services Digital Net-
  work), 232
ISO (Internal Standards Organization),
  232

K

Kilometer, 232
KPSI, 232

L

Labeling the wiring system, 177
LAN (Local Area Network), 232
LAN multistory, 156
Laser, 232

Layout drawings, 225
LED (Light-emitting diode), 232
Lightning arrestor, 36
Link and connect panel, 195, 199
Load, 1
Local area network (LAN), 94, 232
Local exchange, 232
Local room cable panel, 194
Loop, 232

M

Main distribution frame, 36
Main frame, 232
Management information system, 183
Management problems:
  cost, 204
  database development, 204
  tracking, 204
  wiring, 202
Master panel, 189
Mechanical splicing, 232
Mechanical supports, 116
Megahertz (MHz), 233
Metric prefixes, 233, 238
MFD (Main frame distribution), 232
MICROM Systems, Inc., 233
Micropresser, 232
Minibundle cable, 233
Mode, 233
Modem, 233
Modulation, 29, 233
Molding strip, 144
Monitor, 233
Multichannel cable, 233
Multidrop networks, 95
Multifiber cable, 233
Multimode-fiber cable, 75, 233
Multiplexing, 29, 233
Multipoint networks, 95, 233
MUX, 233

N

NEC (National Electrical Code), 233
Network, 233
Network access protocol, 102
Network connections, 95
Network diagnostic tools, 113

Network topologies, 94
Networks:
  access methods, 96
  bus, 96
  hybrid, 99
  multidrop, 95
  multipoint, 95
  ring, 102
  star, 99
  tree, 100
Noise, 233

O

Octal, 234
Ohm's law, 4
Open Systems Interconnections
  (OSI), 102
Optical fiber, 234
OSI functions, 104
OSI protocol, 159
OTDR (optical time domain reflectrome-
  ter), 234
Output device, 234
Outside cable, 136
Outside consultants, 107
Overhead wiring, 137

P

Passwords, 185
Patch panel, 67
Phase modulation, 30, 234
Pigtail, 234
Plenum, 234
Plenum cable, 234
Point-to-point connection, 96, 234
Port, 234
Power loss, 6
Power requirements, 110
Powers of ten, 238
Protocol, 234
Public branch exchange (PBX), 101, 234
Pulse spreading, 79
Pulse:
  fall time, 13
  rise time, 13
Punch-down blocks, 149
Punch-down tool, 45

PUR, 234
PVC, 234
Pythagorean theorem, 11

Q

Quality control, 122, 225

R

Rack inventory chart, 219
Raised floor, 141:
  cells, 141
  conduit, 142
Receiver, 235
Red flag alert, 160
Reference point to power level, 172
Refraction index, 235
Repeater, 102, 235
Request for bid, 212
Request for bid proposal (RFP), 126
Resistance, 4
Retrofitting, 204
Ribbon cable, 79
RING, 235
Riser, 144, 235
Rise time, 13, 235
Room cabling information, 118
RS-232 connector, 42, 235
RS-232 interface, 224

S

Search inquiries, 186
Security 73, 92
Service availability panel, 196
Service desk, 124
Service impact, 122
Shielding, 26, 28
Shielding effectiveness vs. frequency, 45
Short test, 163
Signal loss, 5
Signal-to-noise ratio, 6, 169, 171, 235
Single-mode fiber, 75, 235
SNA fiber cable connectors, 84
SNA network protocol, 104
Software, 235
Source, 235
ST fiber cable connectors, 84

Standard test procedures (STP), 167
Star network, 97, 235
Station, 235
Station equipment, 110
Step-index fiber, 75, 235
Strength member, 77
Structured query language, 185
Support personnel, 112
Support room requirements, 112
Symbols, 240
System configuration worksheet,
    181, 220

T

Tap, 236
Technical support, 107
Technical support group, 175
Telco, 236
Telephone service, 120
Teleprocessing, 236
Terminal block, 236
Termination, 236
Test equipment requirements, 112
Test instruments:
    inductive amplifier, 167
    optical fault finder, 171
    time-domain reflectometer, 170
    tone, 167
Testing, 225:
    coaxial cable, 164
    continuity, 161
    fiber-optic cable, 169
    twisted-pair, 161
    short test, 163
Test-tone instrument, 167
Tight buffer, 236
Time division multiplexing (TDM), 29
Time-domain reflectometer, 170
TLP system measurements, 174
TNC, 236
Token ring, 236
Tools: twisted pair, 43
Tracking cable systems, 204
Training, 112
Training time, 236
Transmitter, 236
Transmission media:

shielding, 26
twisted pair, 23
Tree network, 100, 103
Troubleshooting, 158
Tunnel, 145
Twin-axial cable, 28, 41
    applications, 32
Twisted pair:
    advantages, 46
    application, 28, 31
    bit rate vs. length, 40
    cross talk, 25
    current vs. conductor size, 40
    dB vs. frequency, 42
    disadvantages, 47
    frequency vs. attenuation, 41
    grounding, 37
    rise time vs. distance, 42
    termination, 31
    wire size, 26
Two-piece coaxial connectors, 62

U

Under carpet cable, 79
Underground cable, 136, 146
UL (Underwriters Laboratories, Inc.),
    236
Up time, 236
User population, 108

V

Vendor warranty, 225, 226
Vendors, 107
Vertical riser, 145
Voltage, 236
VOM, 161

W

Warranty, 225
Wavelength, 80, 236
Wire cable grip, 140
Wire cutter/stripper, 46
Wire mold, 144
Wire resistance table, 165
Wire stripper, 46
Wire stripper/crimper, 46

Wire table, 26
Wiring closet worksheets, 180
Wiring installation:
  air conditioning, 115
  bidding plans, 111
  bid proposal, 126
  cable requirements, 111
  chilled water, 115
  data base, 108
  documentation, 108
  enviromental, 114
  existing cable, 107
  grounding, 115
  inventory, 118
  labeling, 128
  phone, 110
  planning, 106
  plan preview, 123
  power allocations, 110
  project scope, 107
  safety, 129
  scheduling, 126
  service desk, 124
  service impact, 122
  station equipment, 110
  support room, 112
  test equipment training, 112
  work areas, 108
Wiring labeling, 128
Wiring plan, 131
Wiring resistance, 5
Wiring rules, 135
Wiring tunnel, 145
Work area inventory sheets, 179
Work areas, 108, 236
Work assignments, 223
Work order panel, 197
Worksheets:
  controller room, 180
  system configuration, 181
Wiring closet, 180
Work stations, 237
Writing a bid proposal request,
  205, 207

Z

Zero transmission level, 173